Theology, Disability and the New Genetics

Why Science Needs the Church

Theology, Disability and the New Genetics

Why Science Needs the Church

Edited by
John Swinton and Brian Brock

t & t clark

Published by T&T Clark
A Continuum imprint
The Tower Building, 11 York Road, London SE1 7NX
80 Maiden Lane, Suite 704, New York, NY 10038

www.continuumbooks.com

All rights reserved. No part of this publication may be reproduced or transmitted in any form or by any means, electronic or mechanical, including photocopying, recording or any information storage or retrieval system, without permission in writing from the publishers.

Chapter 6 originally published in the *Florida State Law Review*: 'Aren't we all Eugenicists?: Commentary on Paul Lombardo's "Taking Genetics Seriously"' (30.2 [Winter 2003]: 223–34).

Chapter 4, 'Arguing about Genetics and Disability', copyright © Tom Shakespeare.

Copyright © John Swinton and Brian Brock and contributors, 2007

British Library Cataloguing-in-Publication Data
A catalogue record for this book is available from the British Library

Typeset by Free Range Book Design *&* Production Limited
Printed on acid-free paper in Great Britain by Athenaeum Press Ltd, Gateshead, Tyne and Wear

ISBN-10: HB: 0-567-04543-9
          PB: 0-567-04558-7
ISBN-13: HB: 978-0-567-04543-0
          PB: 978-0-567-04558-4

Contents

Abbreviations vii

Preface ix
John Swinton

Introduction: Re-imagining Genetics and Disability 1
John Swinton

Part 1 Disability and the New Genetics: Experiencing Disability 27

1. Being Disabled in the New World of Genetic Testing: A Snapshot of Shifting Landscapes 29
   Brian Brock and Stephanie Brock
2. 'What's Wrong with You?' Disability and Genes as Ethics 44
   Christopher Newell
3. Life as Being in Relationship: Moving beyond a Deficiency-orientated View of Human Life 57
   Martina Holder-Franz
4. Arguing about Genetics and Disability 67
   Tom Shakespeare

Part 2 Eugenics and the New Genetics 73

5. To Form a More Perfect Union: Mainline Protestantism and the Popularization of Eugenics 75
   Amy Laura Hall
6. Aren't We All Eugenicists Anyway? 96
   Mary B. Mahowald

Contents

Part 3 The Promise of the New Genetics   115

7   *Conditio Humana* as Viewed by a Geneticist   117
    Walter Doerfler

8   Researching Genetics and Health: Implications for Public Health
    Primary Care Medicine   132
    Blair Smith

9   Genetics, Conversation and Conversion: A Discourse at the Interface
    of Molecular Biology and Christian Ethics   146
    Brian Brock, Walter Doerfler and Hans Ulrich

Part 4 Theological Reflections on the New Genetics   161

10 Life's Goodness: On Disability, Genetics and 'Choice'   163
    Hans S. Reinders

11 Angels with Clipped Wings: The Disabled as Key to the Recognition
    of Personhood   182
    Bernd Wannenwetsch

12 Disability and the Quest for Perfection: A Moral and Theological
    Inquiry   201
    Brent Waters

13 The Broken Body and the Disabled Body: Reflections on Disability
    and the Objects of Medicine   214
    Jeffrey P. Bishop

Conclusion: Fragility and Grace; Theology and Disability   234
Robert Song

List of Contributors   245
Index   247

Abbreviations

AES American Eugenics
ERO Eugenics Record Office
IVF *in vitro* fertilization
PBS Public Broadcasting System
PGD pre-implantation genetic diagnosis
YMCA Young Men's Christian Association
YWCA Young Women's Christian Association

Preface

The origins of this book

This book is the product of a unique, international multidisciplinary symposium that took place in May 2005 in the Centre for Spirituality, Health and Disability at the University of Aberdeen.[1] The symposium sought to explore the practical and theological implications of recent developments in genetic science for the lives of people with disabilities. It drew together a team of theologians, ethicists, doctors, scientists and people with disabilities from the United Kingdom, Germany, Holland, Switzerland, Sweden, Europe, Australia and the United States. The object of the meeting was to create a working community of scholars from different backgrounds and with different perspectives and put together an original text that would open up new areas for dialogue and development around issues relating to genetics and human disability. Over a three-day period this team worked together, laying the foundations for this book. Each paper was read and critiqued prior to the meeting, and discussed and reflected on critically by each member of the group during the two days together. In this way each of the chapters presented in this book are genuinely multidisciplinary, containing input and perspectives from all of the disciplines represented at the symposium.

The focus on disability

The particular focus of this book is on disability. Some of the symposium's participants had been sensitized to key issues through their own experiences of disability, others by disability within their families, and others through personal or professional engagement with people who experience disabilities. Experiences such as these alert people to the importance of asking critical questions of genetic technology.

The meeting adopted an unashamedly theological focus. Participants had noticed that the loudest voices within the debate surrounding genetic science and genetic technology have emerged from the complex narratives of bioethics, economics, politics, medicine and science. The questions that participants wished to ask of genetics emerged from a different perspective, that of theology and the experience of disability. What does genetics and genetic technology look like when it is perceived and questioned from these perspectives? What does genetic technology look like if we begin with the assumption that God is real and active in the world and that disability does not have to be understood primarily in terms of suffering or tragedy? What moral shape does genetic technology take on when the discussion begins with the assumption that people with even the most profound forms of disability live lives that are worth living, lives that have deep significance, meaning, and godly possibility? How does such a perspective affect our perception of the situation of people with disabilities and the response of individuals, communities and societies to their presence? It is as we struggled to answer questions such as these that this book came into existence.

As with all such projects we have many people to thank. Thanks to Iain Torrance, the president of Princeton Theological Seminary who, before he left the University of Aberdeen, provided us with the finances that underpinned our meetings. Thanks to Lindsay Carey and Susanne Rappman for their contributions to our meetings. Thanks also to Tom Shakespeare allowing us to reprint his paper in this text and to the Florida State Law Revue for permission to reproduce Mary Mahowald's paper. We used these two papers to supplement the symposium discussions and we're delighted to include them here. We are extremely grateful to Stephanie Brock for her considerable editing skills, which she put to very good use during the process of constructing this text. There are of course many other people who have influenced this book and helped to shape our thinking and practices. I would particularly like to thank all of the people with disabilities with whom I have worked over the years who have opened my eyes to the importance of disability as a social, personal and theological hermeneutic. The blessings you have given to me are inexpressible. Finally, thanks and glory to God for enabling all of us to see things a little more clearly, and in seeing clearly helping us to live a little more faithfully.

John Swinton and Brian Brock
University of Aberdeen

Note

1. Centre for the Spirituality, Health and Disability, Aberdeen University, www.abdn.ac.uk/cshad

Introduction: Re-imagining Genetics and Disability

John Swinton

In a culture that has learned well how to imagine how to make sense of a world without reference to the God of the Bible, it is the preacher's primal responsibility to invite and empower and equip the community to reimagine the world as though Yahweh were the key and decisive player.

(Walter Brueggemann)[1]

I think there will be a change in our philosophical understanding of ourselves . . . Three billion bases [of a human DNA sequence] can be put on a single compact disc (CD), and one will be able to pull a CD out of one's pocket and say, 'Here's a human being; it's me!'

(Walter Gilbert)[2]

Scientists should find the gene that makes people pick on those who are different. Then our lives would be better.

(Person with a learning disability)[3]

In this introductory chapter I will highlight some of the key themes that run throughout this book and begin to set a context for the various discussions and perspectives presented by the authors in this book. It is not my intention to attempt to summarize the complexities of each argument. Rather, I want to capture some central themes and offer a perspective and some provisional theological understandings that will guide the reader into the areas that are developed more fully as the book progresses. I write this chapter as an academic who has spent most of his life working alongside people with disabilities; previously as a nurse, then as a chaplain, and now as a researcher whose work emphasizes the participatory nature of research and the necessity of doing research *with* people whose life experiences include disability rather than *on* them.[4] It is in such a spirit of partnership and collaboration with people whose life experiences include disability that I offer the insights presented in this chapter.

The 'problem of disability'

Despite the public rhetoric of inclusion, equality and citizenship for people with disabilities, there remains a strange silence around the implications of genetic science and technology for the lives of people with disabilities. One of the intentions of this book is to begin to listen to the meaning of this silence and explore what it might mean to name it. The rise of genetic technology, particularly prenatal screening, has led to a concomitant decrease in the birth of people with disabilities. Many within our society assume this to be a good thing: the writers of this book are not so sure. Linda Ward, in reflecting on the use of prenatal screening, notes that

> The consequences of advances in genetic knowledge and the huge proliferation of prenatal tests, has not . . . been therapy, treatment or 'cure' for a fetus detected as having an impairment; the anticipated outcome of a positive prenatal test for impairment remains abortion. Hardly surprising then, that many in the disability community, and their supporters, are deeply concerned that societal acceptance, even welcoming, of increased genetic testing signals powerful messages about disabled peoples' fundamental right 'to be'.[5]

Ward's point is an important one. Rather than offering treatments and cures, a good deal of genetic technology has become focused on identifying disabilities and seeking to prevent them, with prevention primarily defined in terms of abortion. Ward suggests that

> The lack of explicit, public acknowledgement that the outcome of increased prenatal screening and testing is an increase in abortion on the grounds of fetal impairment has eased the rapid growth and routinization of prenatal testing, without concurrent public debate on the two issues most centrally involved: abortion and disability. Fundamental issues and assumptions implicit in routine practices in this area are rarely surfaced and reviewed.[6]

One of the tasks of this book is to begin the work of surfacing and reviewing these fundamental issues and making explicit that which is implicit.

Genetic variation and the spectre of eugenics

The use of genetic technology to prevent the birth of disabled children raises the question of why we want to live in a society within which a person's desirability is determined primarily by the configuration of their genes and the vacillating social consensus as to whether or not we should accept or reject that particular configuration. Somewhat ironically, one of the things that the Human Genome Project has taught us is that there is no such thing as *the* human genome. Allen Verhey puts this point well

> What is it that the Human Genome Project maps? Not the human person. Not the human body. Not even that thing called the 'human genome.'

Introduction: Re-imagining Genetics and Disability

There is no such *thing* as '*the* human genome.' The Human Genome Project itself has reminded us that genes differ from person to person. The aim of the project was to publish the average or 'consensus' sequence of two hundred different people. But that will provide a map neither of everyone nor of anyone. Does the 'human genome' have blood group A? Or B? Or AB? Or O? We know where to look on chromosome 9 for a marker for blood type, but even if we look carefully, we will not see the blood type of 'the human genome.' We will rather see that variation is an inherent and integral part of the human genome.[7]

The Human Genome project tells us nothing about that which is most meaningful to individuals: who am I? Who do I want to be? What are my hopes, my dreams, my expectations, my loves? It tells us nothing about bodies and their histories. It informs us not one bit about what it means for those bodies to have been created by a God who loves them and desires them to be loved and to learn what it means to love others irrespective of particular disabilities.

Despite claims that we have discovered the universal key to the human self (as Harvard geneticist Walter Gilbert suggests in this chapter's epigraph),[8] genetic configurations are in fact much more flexible and varied than we are led to believe. If genetic variation is in some sense the norm, why is it that some forms of genetic configuration are considered to be more acceptable than others? Is it simply the desire to avoid suffering, or could there be another dynamic at work?

The question of eugenics

Asking such questions inevitably brings us to the issue of eugenics. Simply put, eugenics is a form of social theory that advocates the improvement of human hereditary traits through the use of various forms of intervention. We can divide eugenics into two main categories: positive and negative. *Positive eugenics* relates to the selection of particular traits that are considered desirable, applying techniques and interventions to ensure that these traits are developed and negative traits eliminated. *Negative eugenics* has to do with the elimination of traits or in its extreme form of individuals or groups whose genetic or racial makeup is considered inferior or undesirable. Amy Laura Hall brings to our attention in Chapter 5 the subtle (and not so subtle!) ways in which eugenics has historically involved a quest for the 'creation' of people who are healthier, more intelligent, more beautiful or more economically viable than others. This has often involved both Church and society in deeply questionable social practices, including preventing the birth of or even eradicating disabled people. The depth of the dangers of eugenic thinking was revealed paradigmatically in the atrocities of the Nazi regime, wherein many thousands of disabled people lost their lives during the Holocaust for no other reason than that they were disabled, i.e. their biological or genetic makeup was considered inferior and undesirable. The social power of eugenics was also revealed clearly in the various programmes of enforced sterilization aimed at people with disabilities that were common policy in Europe and the United States until relatively recently.⁹

None of the writers whose work is presented in this book would equate genetic technology with the types of atrocity perpetrated by the Nazi regime. As Tom Shakespeare points out in Chapter 4, the main difference between the eugenics of the Nazis and the present practice of prenatal screening is that 'pre-1945 it was a matter of state policy and it often involved coercion. In present-day Western countries, pregnancy termination is the free choice of individual women and men, within the parameters of the law.' This sets the discussion about eugenics and contemporary genetics in a quite different parameter from the negative of the Nazi regime. Nevertheless, the spectre of eugenics remains albeit of a different order. In their chapters in this book, Hall, Shakespeare, and Mahowald bring the question of eugenics in different but complementary ways into the current debate around genetics and disabilities, and show clearly the ways in which individuals and societies can knowingly or unknowingly be drawn into participating in or condoning eugenic practices. When this happens, people with disabilities find themselves in a very vulnerable position.

Disability and the new genetics: seeing beyond the 'obvious'

It is clear, then, that while genetic technology can be healing and hopeful for some members of our society, for others the same technology can be threatening and indeed deadly. Left unchallenged, as Christopher Newell points out clearly in Chapter 2, genetic technology can become a conduit through which oppressive attitudes and values easily become embodied in practices that are eugenic in nature if not necessarily in intent. As a person with a disability Newell has lived experience of the consequences of such attitudes, experience which he uses to good effect to show the importance of clear ethical thinking in relation to genetics and disability.

At the heart of the issue lies the social contradiction between the expressed intention of genetic technology and the actual consequences of its application. Take, for example, the practice of prenatal testing and in particular amniocentesis. Amniocentesis is a form of genetic testing which focuses on identifying particular genetic variations, primarily Down's syndrome, with a view to allowing parents to choose whether or not the child should be born. At the moment, most abortions that take place due to a positive result from amniocentesis are children diagnosed as having Down's syndrome.¹⁰ The expressed reason for this is that the birth of such a child would cause distress and suffering. But whose suffering is actually avoided by identifying and aborting a child with this particular genetic configuration? It is clear that many people with Down's syndrome live very full and happy lives. Even those who have physical impairments which can be argued to cause suffering do not necessarily suffer, at least not in the ways we assume that they do. Take for example this statement from a young woman with Down's syndrome: 'I was born with a hole in my heart. When I was little it needed a patch and I was very ill. It might be because of this but I have always felt special ... God is my best friend. God made me special because I was special to him.'[11] The standard 'narrative of suffering' would see this woman's situation as 'proof' that it is better to eliminate Down's syndrome in order to avoid unnecessary suffering. Not only does she have a learning disability, she also has a serious heart condition. 'What kind of God would allow such suffering? It seems so unfair. Surely we are justified in doing all that we can to eliminate it?' And yet, when we listen to the woman's own interpretation of her experience, something quite different emerges. Rather than perceiving herself as facing a life of meaningless suffering, her disability seems in fact to have drawn her closer to God and enabled her to develop a positive self-identity and a strong spiritual relationship. One narrative tells of the perceived inevitability of suffering and the simple 'fact' that the world would be a better place without her; the other narrative relates the story of a young woman who has a rich spiritual life and who feels 'special' *because* of her disability, not in spite of it. My point here is not that some people with cognitive disabilities such as Down's syndrome do not suffer: some do and others don't. The point is that the quality of a person's life cannot be anticipated by means of a genetic test. And yet that is precisely what we do.

This of course raises important issues for people with this form of genetic variation. How must the general acceptance of the use of genetic technology designed to eliminate Down's syndrome impact upon the lives of people with Down's syndrome who are actually with us? The small amount of research that has been done into the opinions of people with learning disabilities about amniocentesis has indicated that people are shocked that the question, 'should a child with Down's syndrome be aborted?' is even on the social agenda.[12] Why? Because their experience is that people with this form of genetic disability lead lives that are valued and considered by themselves, their families and their friends to be worth living.

On reflection, the majority of suffering that people with Down's syndrome experience is primarily inflicted upon them by the ways in which society rejects them and chooses not to value their life experience. As one person with a learning disability put it: 'The foetus should be aborted if a test shows it has a learning disability because I don't think it should be born into a cruel world'.[13] Such a statement captures something of the seriousness with which people with learning disabilities treat the stigmatizing assumptions and derogatory attitudes that society presents to them. The ready acceptance by society of tests designed to identify and eliminate people who share their life experiences can only be deeply unsettling.

It must be emphasized that the moral and relational shape of our society is not accidental. It is a *choice*, rather than a fixed, unchangeable reality. In a society with a different moral system that did not depend on competitiveness, individuality and productivity, the concept of Down's syndrome, and indeed the concept of intellectual disability in general, simply would not exist. In other words, the very terms 'Down's syndrome' and 'intellectual disability' indicate adherence to a particular moral code and system of valuing human beings that reflects the ideals of individualism, liberalism and a capitalist economy. Within a society which uses the criteria of independence, productivity, intellectual prowess and social position to judge the value of human beings, people with physical, psychological or intellectual disabilities will necessarily be excluded and downgraded as human beings of lesser worth and value. If this is so, it becomes clear that society is, at least partially, responsible for the disablement of people with Down's syndrome.

We refuse to accept Down's syndrome as a valid and acceptable genetic variation and way of living a human life, not because it is inherently problematic for those who have that life experience but because of the ways in which we have *chosen* to build our communities and the types of values and moral frameworks we *choose* to make our norms. This is not to take away from the real social, economic, emotional and relational issues that people with Down's syndrome and their families may face. Martina Holder Franz explores in detail in Chapter 3 the complex and difficult pastoral issues surrounding accepting and bringing up a child with a disability. Such parenting is not an easy task. It is, however, to raise our consciousness to the fact that there may be more going on in the socially acceptable and apparently compassionate act of amniocentesis than our cultural sensitivities allow us to notice.

We love you . . . now that you are here . . .

All of this alerts us to a significant and quite profound social contradiction. Major strides are being made towards affirming the full personhood and humanity of people with disabilities and safeguarding their rights, value and personhood in law.[14] At the same time, it is also legally, and for many morally, acceptable that certain forms of disability should be prevented from coming into existence through the use of genetic screening to prevent the birth of disabled children. We therefore have an odd situation wherein there is a public discourse that is pro-disabled people and wary of eugenics and eugenic intentions, and a healthcare context within which eugenic activities aimed at people with disabilities appears to be accepted and acceptable. What kind of message does this state of affairs present to people with disabilities? On the one hand, we claim we want to accept and welcome people with disabilities *when they are here*, and on the other we say, 'But everyone would be better off if you were not here at all'.[15]

This state of affairs sends out a particularly negative message to those people currently living with the conditions that we are trying to eliminate. We allow disabled children to be born because we think that their lives are worth living. We prevent them from being born because we do not. As Nelson correctly points out:

> A decision to abort based on the fact that the child is going to have specific individual characteristics such as mental retardation, or in the case of cystic fibrosis a build-up of mucus in the lungs, says that those characteristics take precedence over living itself, that they are so important

and so negative, that they overpower any positive qualities there might be in being alive.[16]

This line of thinking has come to be known as the expressivist argument. It suggests that the idea of identifying unborn children who would be disabled and choosing to abort them because of this identification sends a negative message to those people who have been born and who share the 'negative' characteristics of the aborted child. The message is that their lives are not really worth living, or if they are, they are certainly inferior human beings. What is perhaps most troubling is that, unlike other forms of stigma that are primarily social and psychological (i.e. in principle changeable), this form of stigmatization is ontological. People with such genetic disorders are perceived as flawed in their most fundamental component: their DNA. In a very real sense 'they are not like us' and can never be like 'us', no matter how much 'we' change 'our' attitudes and shift 'our' values. It is then very easy for 'us' to make decisions about 'them' that we would not make about the children with whom we have a more similar genetic inheritance.

But, of course, 'they' are like 'us' . . . indeed they *are* us! 'They' are not strangers but potential friends. The problem with prenatal testing for disability is that it tells us nothing about the person as person. It doesn't tell us about who they are, who they will become or what kind of friend they will be; it tells us nothing about what it might mean to love a child whose genes vary from our own; it provides us with no information about the type of community that we need to become to welcome as friends people perceived to be different. Instead it highlights some potential aspects of the individual person that will have varying degrees of significance within their life. As one woman with a genetic disability observes:

> I know that amniocentesis can't tell any parents what kind of child they will have. It can only tell what disability might exist in that child. Amniocentesis could never have told my mother that I would have artistic talent, a high intellectual capacity, a sharp wit and an outgoing personality. The last thing amniocentesis would tell her is that I could be physically attractive.[17]

By highlighting aspects of the unborn child which are perceived by society as negative, such technology turns the attention of individuals, families, communities and society away from the significance of the person-as-person, and what they have and bring which is positive and life-enhancing, towards the person-as-abnormality. One person with a learning disability puts this point well:

> There should be tests of women who are pregnant to see how the baby is. If it has Down's syndrome, the parents need someone to talk to. *They need to find out what people with Down's syndrome can do.* You should think of the baby as a baby first, not just that it has Down's syndrome . . .[18]

The logic of this position is obvious and profound: the social side-effects of genetic technology leave all people who are labelled as 'genetically disabled' with some serious questions about society's perception of their value, worth and dignity.

The question of personhood

It is of course possible to circumvent some of the implications of the critique of the social consequences of genetic technology for people with disabilities by falling back on the language of personhood. When we engage in the types of discussion that have occupied us up to now, so the argument goes, we are not in fact talking about persons. Rather, we are talking about nonpersons:

> 'Foetuses' and 'disabled neonates' cannot be spoken about as if they were actually persons. To use such language is unnecessarily evocative. Such emotionalism is an unhelpful barrier to logical scientific discussion and indeed to the progress of the field of genetic science . . .

And of course this argument has some validity if we define personhood according to a fixed and identifiable series of human abilities, capabilities, attributes and actions. If personhood is defined by whether someone does or does not have certain characteristics, qualities and abilities then the tone and focus of the discussion does inevitably change. But that change is not necessarily for the better.

Personhood, disability and the value of human life

A good example of the problems that this raises for disabled people is found in the work of the ethicist Peter Singer. He defines personhood as a person's capacity for self-awareness, self-control, a sense of the future, a sense of the past, the capacity to relate to others, concern for others, communication and curiosity.[19] Not to have these attributes means that one remains genetically human but cannot be classified as a person. Nonpersons are entitled to receive the moral protection that the label 'person' ascribes upon its bearer. Singer's position emerges from a basis in preference utilitarianism. From this perspective, utility is defined in terms of preference satisfaction. Preference utilitarians argue that what is right in any given situation is defined by that which produces the best consequences. For Singer this relates to the maximum amount of happiness that emerges from the consequences of an ethical decision. Singer uses this premise to justify the euthanasia of disabled infants before and after birth. He argues that disabled infants (nonpersons) can be killed up to 28 days *after* their birth if they are deemed by medics and parents not to bring maximum happiness to their families.[20] If the child does not bring the parents a maximum amount of happiness, and if it is not considered to be a person for the reasons highlighted above, then the parents are under no moral obligation to protect or sustain the life of the child. In Singer's opinion, this is what we do anyway through the types of technologies that we have already discussed in this chapter. The only difference is that his time-scale is a little later than is currently acceptable.[21]

Defining personhood according to a utilitarian understanding of a human being's abilities or inabilities leaves the essence of moral life and decision-making to the vicissitudes of human happiness. Thus, 'foetuses', 'defective neonates', people with Alzheimer's disease and new-born children find themselves stranded and unwillingly bound to a world of 'its' and 'things': a world which is profoundly shaped by our schooling in consumerist assumptions.[22] Those considered to be nonpersons find themselves defined in terms that effectively turn them into commodities. We do not view *persons* as commodities that can be accepted, rejected or exchanged for better models. Persons are fellow creatures whom we should care for and seek ways of being with. *Nonpersons*, however, can easily be viewed as commodities to whom we owe no moral or relational obligation. Commodities by definition exist to make us happy: they are expendable, simply matters of choice. If they cease to fulfil this role, then we discard them and move on to the next commodity that will (we hope) bring us even more happiness, such as another healthy child. Of course this is rarely the case, as the rising rates of anxiety and depression throughout the consumer-driven Western world would indicate. Nevertheless, we continue to desire happiness and to look to commodities as the means to such an end.

Quality control

Commodities do not require love, acceptance or hospitality. Instead, they are subject to what Brent Waters in Chapter 12 describes as 'quality control'. Waters notices that the idea of 'procreative liberty', prevalent within Western liberal societies, works on the fundamental premise that each individual has a basic foundational right to have or not to have children. Children are not perceived as gifts, but rather are seen as possessions that we have a right to have or not to have. He suggests that within liberal societies individuals who want children have a right to unrestricted access to technologies that will help them fulfil their desire to procreate. Since all competent adults have a right to obtain a child, they also have the right to obtain one that is desirable to them, otherwise why would one want to hold on to a commodity that might not meet one's expectations? Consequently, Waters argues, quality control has become a central aspect of procreative liberty and, we assume, should be respected as such: 'If it is legitimate for parents to want healthy children, then it should be legitimate for them to use both negative and positive techniques to achieve that end'.[23] Those deemed not to meet the often rather arbitrary criterion for personhood are open to the vicissitudes of human rights and choices, quality control and changing and often fickle perceptions of happiness and precisely which types of goods are required to achieve such a state.

Facing up to our choices

The acceptance or rejection of a nonperson and the possibility of deciding on the continuation of their existence or the initiation of their nonexistence depends almost entirely on the particular desires, hopes or expectations that individuals and societies place upon them. The language of 'foetus' and 'defective neonate' (Allen Verhey describes such language as neologisms designed 'to help us forget that it is our children we are talking about') allows us to commodify our children. In this unhelpful way, the concept of personhood protects us from the impact and unpleasant realities of the hard decisions and choices that we sometimes have to make. By removing the unborn child with a disability from the world of persons, it is fairly easy to make decisions about whether or not we choose for it to be born or how we treat it when it is born. If personhood is defined according to a list of attributes that creatures categorized under the genus *Homo sapiens* must have before they are able to achieve, retain and continue to retain over time the designation of 'person', then people with various forms of disability find themselves in a most precarious situation.

Of course, maintaining such an attributes-based understanding of personhood is often easier to do in abstraction than in reality, even for its strongest advocates. The contradiction in Peter Singer's work, revealed by his care and concern for his mother who experienced Alzheimer's disease (a neurological condition which would inevitably end in her becoming a nonperson), and contrasting sharply with his earlier theoretical works that argued forcefully that people who lose their attributes to dementia should be killed indicates a dissonance that is telling.[24] Perhaps the reality of love throws out a practical challenge to the theory of personhood that is hard to resist. Is there more to family and a shared family history than mere logic and rational argument are capable of dealing with? Perhaps personhood relates not to what we can do, but to whose we are?

Personhood, family and the significance of being

I think that Hans Reinders in Chapter 10 is correct when he suggests that we need to take more seriously the providence of God and the moral and theological significance of simply being. In a world where 'doing' reigns supreme in terms of a person's moral and social standing, Reinders calls us back to remember the significance of being. He argues that we need to move on from commodifying attitudes and begin to realize the profound theological significance in simply being and reflecting on the nature of being as it relates to God's providential care for the world. Reinders draws out some of the implications of this suggestion and works us through the implications of the providence of God for our understanding of disability. In concluding that 'life is good because it is what it is, not because we chose it, or would have chosen it had there been a choice', Reinders provides a powerful theological challenge to the implicit and explicit power of commodification that permeates the various contemporary and historical discussions on personhood.

Aspects of Reinder's argument call us back to the Christian contemplative tradition, wherein personhood is revealed not through human attributes and abilities to act in particular ways but simply in being with God; resting in God,[25] attending to God as God attends to us. Such contemplative being is not an act of will or an exercise of human intellect, wisdom or reason. Rather, it is a resting in God in the power of the Holy Spirit which transcends human ability and capabilities. This tradition teaches us the importance of loving God for God's sake, not for what God will do for us or what we might want God to do, but simply for who and what God is. This is important because to love in this way is precisely how we all hope and pray that God loves us. Contemplation does not require anything other than being with God, attending to God and allowing God to attend to us. Such mutual attention is not based on cognitive or physical abilities. It is a gift of the Spirit that brings a mutual valuing and recognition of the other and a relational movement to actualize that value.

There is inherent value in simply being, in accepting and resting in the providence of God. This being allows us to become both the object and the subject of God's love, not because of what we can or cannot do, or because we do or do not have certain attributes, but simply because we are. We are not commodities to be bought and sold, chosen or rejected. We are first and foremost persons, loved and valued just as we are because that is what God is and what God does. Such contemplative resting is not based on choice but on *recognition*. When God is recognized for whom and what He is, there is no choice as to who we will worship. When we learn to recognize God in this way, we learn to recognize others similarly. When this happens we learn to love. Learning to reflect and embody this contemplative way of being holds much potential for an effective response to some of the issues raised thus far.

Reimagining disability and personhood

If we reimagine disability and personhood in the light of such a contemplative perspective, we can begin to understand the nature of personhood quite differently. Scripture is clear that human beings are loved in ways that other aspects of creation and other creatures are not. That love is not extended to human beings because of anything that they have or do. Indeed, it is extended *in spite* of the things they are and do. As Bernd Wannenwetsch correctly points out in Chapter 11, there is something special and unique about being a member of the human species, quite apart from the particular skills and attributes that humans use to actualize and realize the gift of life they are given. Human beings are persons by virtue of the fact that they are human beings, particular objects of God's love and salvific intentions.

If this is so, then the beginning point for all discussion on personhood, and particularly the areas of genetics and disability, is the fact that we are persons by virtue of our Adamic inheritance: we are loved beyond all things by a God who claims us as our Father. *Our personhood is not defined by what we can or cannot do but by whose we are and where we come from.* If our status as persons is aligned according to our membership of the human species, not defined by taxonomy but by a common ancestry with Adam whom God created and described as 'very good', then our perceptions of disability and personhood are inevitably challenged and shifted. We are all members of the human family. This status is unalterable, and not in any sense determined by the presence or absence of disability or certain genetic configurations. It is deeply tied in with the doctrines of creation and redemption. God creates human beings: human beings sin and fall. God is in the midst of redeeming and recreating humanity through Christ's sacrifice. He does this for those who have consciously broken their relationship with God and for whom forgiveness and reconciliation are both available and imperative. It is this lineage and shared history that determines what it means to be a human person.

From the womb to the grave

It is interesting to note that there is evidence within Scripture to suggest that the status of personhood is not confined to a human being's post-uterine experience. What are we to make of the psalmist's reflections on God's involvement with humans even before they are born?

> For you formed my inmost being. You knit me together in my mother's womb.
> I will give thanks to you, for I am fearfully and wonderfully made.
> Your works are wonderful. My soul knows that very well.
> My frame wasn't hidden from you, when I was made in secret,
> woven together in the depths of the earth.
> Your eyes saw my body.
> In your book they were all written, the days that were ordained for me,
> when as yet there were none of them.
>
> (Psalm 139)

This is an important passage for our understanding of genetics and the appropriateness of genetic technology. As my friend and coeditor of this book, Brian Brock, puts it: 'God does not just turn up and install a soul in something humans have made, but is personally involved in human life from the first moment . . . Every child is a unique being created anew and formed by God like Adam.'[26] The psalmist points out that God is involved with the intricacies of human development even before a child is born, and throughout his post-uterine experience:[27]

> It was you who drew me from the womb
> And soothed me on my mother's breast

From birth I was cast upon you
From the womb I have belonged to you.

(Psalm 22)

If this is so, then the idea of discarding an unborn child because of its genetic configuration takes on quite a different light.

Most of us read ourselves into texts like those above. We assume that it is our own 'normality' and perceived 'beauty' that are the primary locus of God's creative activity as it is described by the psalmist. However, passages such as these relate in interesting ways to the doctrine of divine providence and raise some vital questions. If we begin to take seriously the possibility that there may be providential meaning in disabled lives, we may ask why it is that we feel that such passages would *not* apply to someone with Down's syndrome, cystic fybrosis, motor neuron disease, Lysch Nyham syndrome, or Huntington's chorea? Only then would the 'obviousness' of genetic intervention that seeks to eradicate suffering through eliminating persons become less apparent. If God is present, with and for the child who has Down's syndrome as it develops in the womb, in precisely the same way that God is with and for the child society has deemed to be normal, our perception of genetic disability and what a faithful response to people with this experience might be cannot but be shaped in ways which challenge our cultural assumptions.

Preaching without words

Perhaps the most worrying aspect of genetic technology that targets disability is that it totally ignores the fact that God is at work within the lives of people with even the most profound forms of disability. Over the years I have worked with many people whose life experience includes profound cognitive disability. I am always struck by the overwhelming sense that despite the apparent inability of some people to understand and respond in the ways that are expected by the majority, there is much more to their lives than can be seen through eyes which register only pathology and suffering. Even those with the most profound forms of disability are able to worship. Watching people with profound cognitive disabilities worshipping, reaching out and being moved by the sounds of the music and praise opens one up to the fact that God works in ways that far transcend the boundaries of human expectation and comprehension. As the apostle Paul correctly notices: 'In the same way, the Spirit helps us in our weakness. We do not know what we ought to pray for, but the Spirit himself intercedes for us with groans that words cannot express.'[28] The deep groaning of the Spirit becomes apparent and embodied in the gestures of worship that we encounter as we praise together with people whose life experience includes intellectual disabilities. In the context of worship it is clear that these 'undesirable lives' are capable of experiencing God and are open and able to worship him.

As we encounter people with genetic variations and communicational styles which differ from the norm, we are constantly reminded that there is something more to human life than we are often taught by culture. One mother of a profoundly cognitively disabled young man told me the following story about her son:

> When he comes to tell us to look at something out of the window it's really special because he doesn't say very much to us. So the moon is something that he seems to really appreciate and if it's a really bright night and a dark sky and the moon is shining he just has an immense appreciation of that. He wants to stand and look at it for ages and to share that moment of looking at the moon.

This young man, David, rarely communicated, and yet, when he encountered the moon, something shifted in his life. Precisely what, no one knows. Was this a spiritual experience? A sense of awe and wonder? A sense of something beyond? A sense of God's movement? We don't know what it is, but neither do we know what it is not. What we do know is that this deep and meaningful experience of somehow connecting with that which is beyond changed him, and opened him up to something positive and perhaps spiritual, something he didn't experience normally and which many around him didn't consider him to be capable of experiencing. 'With men this is impossible; but with God all things are possible.'[29] For those whose lives include being with people with disabilities such as these, such tender and profound experiences embody the optimism of the psalmist and open all of us up to the possibility that there may be more to the decisions we make about the desirability of disability than we have previously been moved to notice.

This in turn raises more questions about the providence of God, the meaning of disability, theodicy and precisely how Christians should respond to the suggestion that the types of experiences that result from genetic differences may have a deeper meaning than the language of pathology and 'bad genes' alone can express. These are big questions, to which the writers of this book begin to offer some answers.

Why genetic science needs the Church

The discussion presented thus far has made it clear that facilitating a constructive conversation between genetics, theology and disability is difficult, complex and politically sensitive. The evocative nature of the key issues combined with the strange language and complex worldviews of theology and the science of genetics does not make for easy conversation. The wisdom of the gospel frequently sounds like foolishness in the face of the complexities of DNA matrixes, nucleopeptides and recessive phenotypes: terms one struggles to find equivalents for in the pages of Scripture. At first glance the two languages appear to belong to two very different lands. The psalmist's question, 'How can we sing the Lord's song in this strange land?' trips easily from the lips of Christians as they try to find ways of opening up a dialogue. Nevertheless, the initiation of such dialogue is crucial. The question is not *if* but *how*.

Walter Brueggemann, in the epigraph to this chapter, points to the way in which culture has learned to envision itself functioning effectively without the need for God.

> In a culture that has learned well how to imagine how to make sense of a world without reference to the God of the Bible, it is the preacher's primal responsibility to invite and empower and equip the community to reimagine the world as though Yahweh were the key and decisive player.[30]

This is an interesting observation. Reflecting on the types of issues that I have highlighted in this chapter and which are developed throughout this book, it would seem that there is a real sense in which genetic science has learned to imagine itself and to make sense of the human genome without any real need for God. The types of theological issues that have been raised in this chapter are not perspectives that one normally hears as one listens to the public rhetoric of genetic science. This is rather surprising, bearing in mind that genetic science takes place within God's creation, the place where Yahweh rather than science reigns supreme. Perhaps a way forward is for the Church as the Body of Christ to take seriously Brueggemann's invitation to invite, empower and equip that community to reimagine the world with Yahweh as the key and decisive player. Perhaps the dialogical task is to seek ways to re-imagine (use our theological imagination to see beyond that which is 'obvious' towards that which is hidden yet profoundly important) genetic science and technology in the light of the transformative fact that God is the decisive player in all development within the new genetics. When we do this, things begin to look different.

Christ supreme in all things

In his letter to the Christians at Colossae, the apostle Paul stresses the importance of recognizing the supremacy of Christ over all things.

> Christ is the visible image of the invisible God. He existed before God made anything at all and is supreme over all creation. Christ is the one through whom God created everything in heaven and earth. He made the things we can see and the things we can't see – kings, kingdoms, rulers and authorities. Everything has been created through him and for him. He existed before everything else began, and he holds all creation together.[31]

Paul stresses the importance of recognizing the significance of the fact that human beings are creatures; that we live in God's creation and are bound by God's authority. The autonomy, freedom and knowledge that we have are necessarily implicitly or explicitly bounded by this transformative fact. It is true that not all of creation recognizes God as God, and human beings as creatures:

nonetheless, God remains with and for creation now and for eternity. More than that, God in Christ is not simply present alongside creation, but is deeply and actively implicated with it and has authority over it: 'God has put all things under the authority of Christ, and he gave him this authority for the benefit of the church. And the church is his body; it is filled by Christ, who fills everything everywhere with his presence.'[32]

The transformative fact that we are creatures who are inevitably under the authority of God irrespective of whether or not we choose to acknowledge it, sets the discussion of both genetics and disability within a specific frame of reference that is of vital importance for this book. It is God's authority and not the epistemologies of science, pragmatism, economics, politics or disability studies that provides the key facts in discussions about genetics and disability. This being so, the Church has a specific responsibility to recognize and prophetically acknowledge God's will and authority over all things and to act faithfully according to that authority of which it is a benefactor. To say this is not to suggest that the Church owns or has sole access to the authority and will of God: only God is God, and God will use whomsoever it pleases God to use! But the Church does benefit from recognizing the boundaries that such authority places on its practices. The Church's duty is to act on that recognition and to work under that authority in all things and at all times. It is for this reason that genetic science needs the Church. Without the accountability that comes from the recognition of God's authority over all things, genetic science easily strays into ways of practising which, as we have seen, can be deeply problematic for some of the most vulnerable members of our society.

Framed in this way, God's authority is, or at least should be, a necessary moral dynamic within the development of genetic science, and the Church is, or at least should be, actively and prophetically involved in monitoring and shaping its development and application. Viewed in this way, we can no longer allow our responses and strategies to be determined and driven purely by questions of utility (how best can we do this?). Rather, the determining criterion becomes one of faithfulness and obedience. (*Should* we do this even if we can?) This perspective allows us to take seriously human reason and the genuine benefits of the developments within genetic science, but at the same time to view these things through a theological lens that takes seriously the transformative fact that Yahweh is the key and decisive player and that Jesus reigns supreme in the field of genetics as in all things. This perspective enables us to begin to explore the field of genetics as a particular context within which we can encounter God and learn what it means to live under God's authority, and in so doing, have revealed to us a fresh appreciation of God's presence, love and power.

Goodness and mystery

This being so, it is possible to view the developing understanding of the human genome and the practices and technologies that are emerging from it as a place

Introduction: Re-imagining Genetics and Disability

where the goodness and mystery of God are uncovered in ways which may well be deeply theologically challenging, but which hold the potential to be profoundly faith enhancing and transformative. All that God made is good. We need not, in principle, exclude forms of genetic technology from this affirmation if it is shaped by the desire to bring glory to God. Seeking to understand something of the mysteries of human life in illness, suffering and joy is not necessarily problematic. Indeed, when we begin to open up conversations with those who spend their lives working within the strange land of genetics, we very soon discover that the land of genetics is inhabited by many people who love God and who recognize that this form of technology can serve his healing work.

Finding God in the human genome

In Chapters 7 and 8, Walter Doerfler, a geneticist, and Blair Smith, a family practitioner with specific expertise in clinical genetics, show clearly that genetic science and the use of genetic technology is not inherently problematic for Christians. Doerfler correctly points out that theology and genetic science are not necessarily enemies. At many levels they are often pursuing similar goals and asking related fundamental questions: What are human beings? To what extent has the new genetics enabled us to reflect more realistically on the human condition? Doerfler's point is a strong one. Both theology and genetics seek after the good of human beings. Both, in different ways, hold the potential to reveal the magnificence of God and the intricate nature of his creation. Doerfler's chapter presents us with an in-depth overview of the field of genetics and draws our attention in a clear, deep and accessible way to the complexities of the field and the wonders of the human genome.

Similarly, Blair Smith illustrates the healing potential that the new genetics has for the day-to-day practices of medicine. Smith makes a strong case for the importance of genetic technology within primary medical care and highlights its significance for anticipating and dealing effectively with certain common conditions such as heart disease, cancer and diabetes. Smith's perspective opens up the possibility that genetic technology has important potential for enhancing human well-being and for alleviating suffering at a number of levels. A key point for Smith is that genetic science and technology is becoming a central aspect of the way in which medicine is developing in the UK and beyond. We cannot and should not try to turn the clock back. The question is not whether or not Christians talk about genetics, but precisely how they carry out the conversation. While acknowledging that there are problems, Smith argues firmly for the reality that genetics is here to stay. The task of the Christian, Smith argues, is to recognize the real advantages of genetic technology for a significant number of people within society and to begin to explore what types of critical dialogue make most sense as Christians strive to minister faithfully within this area.

In Chapter 9, Hans Ulrich, Walter Doerfler and Brian Brock provide a particularly useful example of just how such a dialogue might work out in practice. As they illustrate well, when it is used faithfully, thoughtfully and prayerfully, the contemporary theory and practice of genetics can be effective in the service of God.

Understood in these ways, genetics can be viewed as being in the service of God; as such, it is bounded by the implications of the underpinning fact that Yahweh is the key and decisive player in this aspect of creation. Studying genetics in this mode should be a prayerful human activity that combines the recognition of the Creator in all things with the necessary responsibility to love others as God loves us and to fulfil Jesus's commission to bring healing and to relieve human suffering.

We are glad that you are here?

Perhaps it is here, as we begin to reflect on the centrality of love for all human activity, that we begin to see what is missing from some of our contemporary practices. As one reflects on the issues that surround disability and the new genetics, one is immediately struck by the lack of a rhetoric of love. Christians believe that the radical heart of God *is* love. Not simply that God loves or is loving, but that God in and of God's self is love. Any serious theological reflection on genetic science *must* begin with the recognition of who God is, and the acknowledgement that the primary purpose of human existence is to love God, self, and others. If Jesus is correct that the greatest commandment is that 'You shall love the Lord your God with all your heart, with all you soul, and with all your mind', then this all-encompassing understanding of love must form the basic practical and hermeneutical principle for all of human life, including our practices within the field of genetic science.

The shape of Christian love is therefore quite precise and particular. God is love, but love is not God! Within Western society the shape of love has become unclear and diverse. As we have seen, in the name of such things as love, compassion, and the alleviation of suffering, people with disabilities can be put in a very dangerous situation. Left to its own ends, the concept of 'love' can lead to practices that are deeply unloving. The love of God the Father as it is revealed in the life, death and resurrection of Jesus is not a means for advancing human happiness; it is not a commodity that can be rejected if it does not meet our needs. Rather, it is gentle, patient, kind and persevering. It is hopeful and protective, rooted in a deep trust in God. Such love is not dependent on anything we can or cannot do. It is not dependent on our love for God, but on God's unending love for us:

> This is how God showed his love among us: He sent his one and only Son into the world that we might live through him. This is love: not that we loved God, but that he loved us and sent his Son as an atoning sacrifice for our sins. Dear friends, since God so loved us, we also ought to love one another.[33]

Love, then, is not an idea, a concept or a feeling; love is a person: God *is* love. Such love does not change its mind about loving because the object of its love has been a disappointment. It is this love that those who claim to know God are called to reveal to the world:

> No one has ever seen God; but if we love one another, God lives in us and his love is made complete in us. We know that we live in him and he in us, because he has given us of his Spirit. And we have seen and testify that the Father has sent his Son to be the Saviour of the world. If anyone acknowledges that Jesus is the Son of God, God lives in him and he in God. And so we know and rely on the love God has for us.[34]

To know God is to act as God does.

Christians bring this love to the field of genetics, and it is according to this love that they are called to offer discernment and guidance. Josef Pieper describes Christian love in this way:

> In every conceivable case love signifies much the same as approval . . . It is a way of turning to him or it and saying, *'It's good that you exist; it's good that you are in this world!'* . . . Human love, therefore, is by its nature and must inevitably be always an imitation and a kind of repetition of this perfected and . . . *creative* love of God.[35]

Love means recognizing and welcoming. In defining love in this way, Pieper draws together the nature of divine love and in an interesting way combines it with the motif of hospitality. Love speaks the words: 'It's good that you exist; it's good that you are in this world', and embodies them in acts of hospitality that reveal the authenticity of that love. This is the essence of faithful discipleship and the heart of worshipful existence. One of the problems with the types of genetic technologies we have been exploring in this chapter, and their accompanying moral and social implications, is that *they are inhospitable and loveless practices*. While the love of God speaks to people with disabilities the words: 'It's good that you exist; it's good that you are in this world', the actions of society produce dissonance and conflict: 'It's not good that you exist; it's not good that you are in this world'. It is more than a little ironic that society seems to welcome genetic technology, being glad that it exists and happy that it is here, in a way it does not offer such love and welcome to the ancestors of Adam who happen to have disabilities. To say to someone that they are not welcome here because they are different or because they will cause us problems is deeply hurtful. To pretend that they are welcome and then to engage in social practices which are designed to ensure that their existence 'won't happen again' is loveless and inhospitable.

Moving on, making a difference

It is one thing to acknowledge God's authority in the field of genetic science; it is quite another actually to change things. Indeed, according to Hans Reinders, the types of change that this chapter has indicated are necessary may not in fact be possible within liberal societies:

> Assuming that disabled people will always be among us, that the proliferation of genetic testing will strengthen the perception that the prevention of disability is a matter of responsible reproductive behaviour, and that society is therefore entitled to hold people personally responsible for having a disabled child, it is not unlikely that political support for the provision of their special needs will erode . . . the question of civic and social hospitality is key, but political liberalism is not ultimately capable of engendering and fostering hospitality towards people with overt, recalcitrant needs. The norms encircling the liberal axis of individual autonomy cannot easily accommodate lives dedicated to the care of perpetually dependent individuals, or admit the intrinsic value of these individuals.[36]

In essence, Reinders suggests that liberal society has neither the practical nor the moral capabilities or desire to protect the disabled and to ensure their future. If this is so, the issue then is cultural rather than political. Hope for the recalcitrant, Amy Laura Hall argues, 'hinges on an abiding account of incarnate life as a gift'.[37] There may well be no political will to make the types of changes indicated by the authors of this book, but change and transformation nevertheless remains a possibility. It will, however, come in a different mode. In Reinders' words:

> The benefits bestowed by love and friendships are consequential rather than conditional, which explains why human life that is constituted by these relationships is appropriately experienced as a gift. A society that accepts responsibility for dependent others such as the mentally disabled will do so because there are sufficient people who accept [this] account as true.[38]

Amy Laura Hall puts this point slightly differently: 'the task of persuading parents to eschew in-vitro genetic diagnosis and embryo disposal, or to refuse ante-natal testing and selective termination, may require at least as much a gesticulation as an argument'.[39] The way of persuasion is to live the types of loving lives within which the option of such technology is no longer there. The Church then is potentially a significant locus of transformation, where people begin to learn what it means to live such lives.

Introduction: Re-imagining Genetics and Disability

Welcoming Adam

I find this line of argument helpful. Perhaps that is why I am drawn to the story of Brian and Stephanie Brock in the opening chapter of this book. In the midst of the complicated theoretical, theological and philosophical debates which surround the areas explored in this book, Brian and Stephanie's reflections on their experiences with their son Adam, who has Down's syndrome, offers a poignant and grounded perspective that points us towards a possible transformative future. By loving and welcoming Adam, the Brocks remind us that the bottom line of all the debates around genetics and disability come down to love. By bringing Adam (as opposed to the facelessness of 'Down's Syndrome') into the discussion, the Brocks refuse to allow the implications of the application of genetic technology to be discussed in abstraction from its loving embodiment in the life of their family. Despite pressures to think and act differently, the Brocks refused to frame their discussion of Adam according to the confines of his genetic structure. From the beginning they were glad that he was here. He was welcomed. He was loved as Adam; not for what he may or may not be able to do but because of who he is . . . Adam, their son.

Having said that, it would be a mistake to assume that somehow Adam justifies his presence in the Brock family through bringing 'transformative insights'. To suggest this would be to suggest that Adam's worth is to be judged on what he can do, in this case, his ability to bring something to the family. The Brocks simply refuse to buy into the idea that what is primarily significant about Adam is his usefulness to others. Adam's worth is not defined by what he enables others to do.[40] As Stanley Hauerwas puts it:

> . . . our attitudes about retarded children [sic] cannot be based on whether or not they enliven certain families. Rather we must ask what kind of families and communities should we be so we could welcome retarded children into our midst regardless of the happy or unhappy consequences they may bring.[41]

This is an important point. In a real sense the Brocks' experience provides an embodiment of the transformative mode of love that Reinders gestures towards and Pieper neatly captures in his definition. They are glad that he is here; they are glad that he exists. The fact that Adam brings transformative insights is a consequence of him being recognized precisely as Adam.

The Brocks' welcoming of Adam does not exclude times when he is less welcome: times when his behaviour is deeply challenging, times when their lack of sleep, their anger and frustration, is far from hospitable. Life with a disabled child can be very difficult. But, importantly for the Brocks, dealing with these issues begins from a position of welcome. Welcoming Adam is not always easy, but the Brocks' patience, trust and timefullness, even when these things are only available in faith, is a sign of hope.

Importantly, the Brocks' story highlights the fact that, in line with the teaching of the psalmists we explored earlier, Adam was welcomed, valued, loved and protected by them *before* he came among them. Despite pressure from the medical profession to take advantage of genetic testing that would have told them Adam was 'disabled' and provided them with the opportunity to refuse to welcome him, the Brocks decided to love Adam and to welcome him. They made a conscious decision to remain engaged and in relationship with Adam irrespective of the nature of his genetic makeup. Adam was not perceived as a faulty commodity that required to be exchanged for a more genetically 'normal child'.[42] Rather, they sought to love and protect him and embody the essence of love.

In focusing on the significance of love, welcome and hospitality for the debates that surround genetics, theology and disability, it is not my intention to try to idealize disability or detract from the fact that people truly do suffer and that families, for various reasons, can be devastated by the presence of disability. There are significant pastoral issues which need to be addressed, a vital observation which Martina Holder-Franz brings out well in Chapter 3 of this volume. Nevertheless, the arguments, narratives and perspectives offered in this book alert us to the fact that we can tell other stories about the experience of disability: stories which challenge some of the basic assumptions that lie behind the justification and use of certain forms of genetic technology; stories which see love and relationship as central rather than peripheral to the discussion. These stories recognize the ways in which genetic science needs the church, and the transformative narrative of the authority of God which the Church is charged to communicate. How we respond to such a challenge will probably illustrate the true level of our commitment to the things we proclaim to be true. Being a critic is relatively easy: becoming an exemplar of a new practical theological paradigm is much more complex and messy.

Structure of this book

In concluding, it might be helpful to offer some pointers as to how the book can be read. There are a number of ways in which one can use this text. The book is intended to allow the reader to begin with the experience of disability before moving on to explore some of the problems and the promises of the new genetics. It concludes with a series of theological reflections on the implications of the new genetics for lives of people with disabilities and the types of communities that we choose to create in response to certain forms of difference. Readers who see the significance of this structure may well wish simply to read the book straight through and follow the flow of the themes and arguments as they are presented. Others may wish to begin by getting a firm grasp on the field of genetics before beginning to think about the more complex theological and philosophical dimensions. The chapters by Doerfler and Smith provide an excellent introduction to that field which may allow some readers to gain a firm grasp of the significance of the arguments developed elsewhere in the book.

If the reader is interested in beginning with the experience of disability, then the chapters by Brian and Stephanie Brock and Christopher Newell

Introduction: Re-imagining Genetics and Disability

will make a good entrance point into some of the central debates that are developed elsewhere in the book. Martina Holder-Franz's chapter exploring some of the pastoral dynamics of genetics and genetic technology again opens up some important experiential dimensions that readers will find useful and challenging.

Amy Laura Hall's work on the history of eugenics offers a challenging social, political and theological perspective, and alongside Mary Mahowald and Tom Shakespeare's reflections on the implications of eugenic thinking for current practice again focuses the reader's attention on the issues discussed elsewhere in the book in a quite specific way. Beginning by thinking about eugenics challenges the reader to look differently at the ways that we accept genetics and some of its practices without a great deal of questioning. This section of the book provides a deep and fascinating overview of the tense relationship between genetics, eugenics and disability.

Readers who are interested primarily in the theology of disability may want to begin with the final section. Here they will find new and innovative theological thinking around key issues of disability, theology and genetics, with important contributions by key theologians Brent Waters, Bernd Wannenwetsch, Hans Reinders, Jeffrey Bishop and Robert Song. This section of the book provides an original foundation for the theology of disability and opens up areas of this field of enquiry that have not previously been explored.

The reader can work with the text in the ways which are most appropriate for his or her needs. However it is used, the writers of this book hope and pray that the chapters they offer will not only challenge, but also transform our practices in ways that enable us to take seriously the demands of love and hospitality, and the transformative fact that Jesus Christ reigns supreme here as in all things. At the risk of distorting the words of Walter Brueggemann which opened this chapter: in a culture that has learned well how to make sense of genetics without reference to the God of the Bible, the chapters in this book seek to invite, empower and equip the community that constitutes the Body of Jesus Christ to re-imagine the world with Yahweh as the key and decisive player.

Notes

1. Walter Brueggemann, *Deep Memory, Exuberant Hope: Contested Truth in a Post-Christian World* (ed. Patrick D. Miller; Minneapolis, MN: Fortress Press, 2000).
2. Walter Gilbert, 'A Vision of the Grail', in D.J. Kelves and L. Hood (eds), *The Code of Codes: Scientific and Social Issues in the Human Genome Project* (Cambridge, MA: Harvard University Press, 1991), p. 96.
3. Linda Ward, 'Whose Right to Choose? The "new" genetics, prenatal testing and People with Learning Disabilities', in *Critical Public Health*, 12.2 (2002), pp. 187–200.
4. John Swinton and Harriet Mowat, *Practical Theology and Qualitative Research* (London: SCM Press, 2006); John Swinton and Elaine Powrie,

Why Are We Here? Meeting the Spiritual Needs of People with Learning Disabilities (London: Mental Health Foundation, 2004).
5. Ward, 'Whose Right to Choose?', p. 188.
6. Ward, 'Whose Right to Choose?', p. 188.
7. Allen Verhey, *Reading the Bible in the Strange World of Medicine* (Grand Rapids, MI: Eerdmans, 2003), p. 150.
8. Gilbert, 'A Vision of the Grail'.
9. Anne Kerr and Tom Shakespeare, *Genetic Politics: From Eugenics to Genome* (Cheltenham: New Claringdon Press, 2002).
10. People with Down's syndrome have a form of genetic variation in which there are three copies of chromosome 21 instead of two.
11. Swinton and Powrie, *Why Are We Here?*
12. Linda Ward, *Considered Choices: The New Genetics, Prenatal Testing and People with Learning Disabilities* (Worcestershire: British Institute of Learning Disabilities, 2001).
13. Ward, 'Whose Right to Choose?'
14. Seen in such examples as the Americans with Disabilities Act (USA) and the Disability Discrimination Act (UK)
15. Hans Reinders, *The Future of the Disabled in Liberal Society: An Ethical Analysis* (Chicago, IL: University of Notre Dame Press, 2001), p. 4.
16. Mary Johnson, 'Aborting Defective Fetuses – What Will It Do?" *Link Disability Journal*, 14 (August–September 1990).
17. Eileen Cronine-Noe, '"Thalidomide Baby" Grows Up', in *Houston Chronicle*, July 1987.
18. Ward, 'Whose Right to Choose?' (emphasis added).
19. Peter Singer, *Practical Ethics* (Cambridge: Cambridge University Press, 2nd edn, 1993), p. 169.
20. Peter Singer and Helga Kuhse, *Should the Baby Live? The Problem of Handicapped Infants* (Oxford: Oxford University Press, 1985).
21. Singer's 28 days should be reflected on in light of the fact that, under the 1967 abortion act in the United Kingdom, a child with a disability can be aborted at *any* point prior to birth.
22. For a deeper critique of Singer's argument see John Swinton, *Raging with Compassion: Pastoral Responses to the Problem of Evil* (Grand Rapids, MI: Eerdmans, 2007).
23. See Ch. 13, p. 206, this volume.
24. Singer, *Practical Ethics*, pp. 175–217.
25. *The Cloud of Unknowing and Other Works* (London: Penguin, 2001).
26. Brian Brock, 'Psalm 22: Standing by God in his Hour of Grieving. Living with the Disabled and the Question of Stem Cells', sermon delivered in St Andrews Cathedral, 14 May 2006. http://abdn.ac.uk/cshad/Brianssermon.shtml
27. Job (10.8–12) states that
    Your hands have framed me and fashioned me altogether;
    Remember, I beg you, that you have fashioned me as clay.
    Haven't you poured me out like milk,

And curdled me like cheese?
You have clothed me with skin and flesh,
And knit me together with bones and sinews.
28. Rom. 8.26.
29. Mt. 19.26.
30. Brueggemann, *Deep Memory, Exuberant Hope*.
31. Col. 1.15–17.
32. Eph. 1.22–23.
33. 1 Jn 4.7–11.
34. 1 Jn 4.12–16.
35. Josef Pieper, *Faith, Hope, Love* (Fort Collins, CO: Ignatius Press, 1997) (emphasis added).
36. Reinders, *The Future of the Disabled in Liberal Society*, p. 14.
37. Amy Laura Hall, 'Public Bioethics and the Gratuity of Life: Joanna Jepson's Witness against Negative Eugenics', in *Studies in Christian Ethics*, 18.1 (2005), p. 16.
38. Reinders, *The Future of the Disabled in Liberal Society*, p. 17.
39. Hall, 'Public Bioethics and the Gratuity of life', p. 16.
40. There does of course appear to be a certain attraction to the idea that people with profound intellectual disability contribute to the good of society by providing insights, perspectives and experiences that can enhance the lives of others, but on reflection it is a highly ambivalent argument. In order to hold on to this argument, as David Pailin correctly points out in *A Gentle Touch: From a Theology of Handicap to a Theology of Human Being* (London: SPCK, 1992), would be to suggest that if humankind were to be wiped out by a plague, and the last tortured survivor were severely intellectually disabled, that person would have no value. By treating a person's worth in terms of units of significance for others, the individual is dehumanized and stripped of the personal significance appropriate to all people for whom God cares, and for whom Jesus died. Instead of allowing the intellectually disabled person's situation to modify and enhance the current system of values, this line of argument in fact attempts to force their situation inside the confines of what is 'known' and 'understood' about the way we value people: it simply will not fit. It sits well with the market-based assumptions to which we consciously or unconsciously subscribe, but it is totally inadequate in terms of understanding the true value of persons.
41. Quoting Stanley Hauerwas in Swinton and Powrie, *Why Are We Here?*
42. It is worth noting that if our reflections on providence and the anthropology of the Psalms that we explored earlier is correct, then the concept of 'normality' is significantly challenged. Normality does not relate to the statistical prevalence of certain traits, but to the continuing presence of God with all human beings and the significance of recognizing that our common ancestry challenges any concept of normality that seeks to exclude certain members of the human family.

Part 1

Disability and the New Genetics: Experiencing Disability

1

Being Disabled in the New World of Genetic Testing: A Snapshot of Shifting Landscapes

Brian Brock and Stephanie Brock

Introduction

This chapter speaks biographically in order to introduce a real-life snapshot of the forces genetic technologies bring to bear on the disabled and their families. We do so as an academic theologian and a neonatal nurse experiencing the joys and frustrations of first-time parenthood. Our son Adam, now two years old, has Down's syndrome, and it is the events of his first six months on which our account draws. We will outline the pressures we experienced as parents of a 'genetically handicapped' child, and then, in conclusion, offer some theological reflections.

The biographical investigation of social phenomenon is related to phenomenology. The Canadian philosopher George Grant explains how such an approach attempts to circumvent the blind spots induced by the experimental methodologies of the social sciences. Grant suggests that some social forces are actually more accessible to a form of observant participation he calls 'enucleation', and his definition of this idea is worth quoting at length.

> In another age, it would have been proper to say that I am attempting to partake in the soul of modernity. When we are intimate with another person we say that we know him. We mean that we partake, however dimly, in some central source from which proceeds all that the other person does or thinks or feels. In that partaking even his casual gestures are recognized. That source was once described as the character of his soul. But knowing, in a strict sense, has excluded the concept of a soul as a superstition, inimical to scientific exactness. To know about human beings is to know about their behaviour and to be able to predict therefrom. But it is not about the multiform predictable behaviours of modern technical society that I wish to write. It is about the animating source from which those behaviours come forth.

What I am not doing is what is done by modern behavioural science, which is not interested in essences. A leading behavioural political scientist, Mr David Easton, said recently: 'We could not have expected the Vietnam War.' This was said by a man whose profession was to think about political behaviour in North America, and whose methods were widely accepted by other scientists. But not to have expected the Vietnam War was not to have known that the chief political animation of the United States is that it is an empire. My use of the world enucleate indicates that I do not wish to use a method that cannot grasp such animations.[1]

Grant's claim is that the most important aspects of our technological age are grasped only through close participatory examination of the contours of experience in technological existence because 'All descriptions or definitions of technique which place it outside ourselves hide from us what it is'.[2] To externalize technique in order to study it is to hide the reality that technique is ourselves, that technology is ourselves in action.

In myriad ways we sustain and perpetuate the social forces that meet in the interactions of medicine with the disabled, interactions which only occasionally emerge into view as something on which we critically reflect. This chapter is one such exploration of what we experienced as the force and direction of our age's soul. Common opinion holds on one hand that the rise of genetic testing is particularly portentous for our relation to the beginning and future of human life, and on the other that genetic technology is an unstoppable expression of what modern society 'is'. Our inchoate perception that we are brushing up against what this age takes to be its fate initiated the processes of critical observation and reflection which follow.

The protest is immediately, and perhaps rightly, raised: 'But we can't simply step outside our age to observe it!' Our response is that observant participation becomes possible only from the vantage points provided by communities which transmit some countercultural logic or impulse and so provide resources for questioning what most take as commonplace. The firm hope in divine presence and the practices of creaturehood learned in Christian worship throw up alternative hopes and alternative ways of being human. This chapter is an attempt publicly to articulate conflicting moments between forms of life in order to make them accessible for moral analysis and public discussion. Such articulation serves state and Church by naming where differing views of reality produce conflicting forms of social behaviour.

Stanley Hauerwas has famously worried that he 'uses' the handicapped for the purposes of just such ethical analysis and cultural critique, as 'canaries in a mine'.[3] Unlike Hauerwas, we raise these questions as directly and permanently involved parties. In an important sense, everyone in society is an involved party, but our family has been involved in a more direct and visible way than most.

Our worry then is not the 'use' of Adam, but co-opting him in his family's and church's political engagement with society in ways that might impoverish him. Receiving this divine gift into our family has produced a wealth of new perceptions of how society (and marriage) works. His very being offers a theological diagnostic value in the social responses he elicits. This has proved especially obvious in our relations with modern medicine. Such enquiry is only part of the process of realizing that how we respond to these forces as parents depends on how we name and respond to the world we inhabit. It is with the hope of more sharply perceiving the soul of our age, and so introducing our son to the life of faith in a broken world, that we think out loud about our lives together. We do so emboldened by the observation that biographical, or first-person experience, combined with sensitive cultural observation and personal conversation, were the only modes of sociological research available to our parents in the faith. In their writings we find many theologically and culturally sensitive conclusions about the spiritual dimensions of the forces their respective societies exerted on them.

Life in genetic limbo

Adam was conceived three months before a planned move from England to Germany. Stephanie wanted to wait until we reached Germany to register the pregnancy: she knew that it would be simpler to begin antenatal care in the country in which Adam would be born. This early disengagement from antenatal medicine left us knowing Adam only in the timeworn ways of experiencing and marvelling at the baby's growth, its first kicks and the changes in Stephanie's body. We believed that a moving and growing baby is a healthy baby, and felt no loss at the lack of medical corroboration of the point. Our conviction was that if the baby stopped moving or growing, then we would have it checked out. As the baby continued moving and growing we felt no pressure to discover if it was not doing so 'normally'. We were happy and all seemed well.

When we arrived in Germany, Stephanie was almost four months pregnant and began the inevitable process of organizing the details of having a child. Not knowing where to start, she first visited the *Frauenklinik* (women's clinic), where she was examined and urged in the strongest terms to have a sonogram as quickly as practicable. The reason for the rush, we learned in time, was a scepticism about Stephanie's stated conception date and a nervousness about nearing the legal cut-off date for an abortion. In the course of discussion it also became clear that the only reason for the sonogram was to see if abortion was indicated. The whole set of considerations were ones with which we disagreed, based on reasons which were not presented up front. Having only gone to the clinic to enquire about antenatal care and birth arrangements, this became our first experience of being offered medical treatment which we neither sought nor desired but were subjected to as a feature of the institutional, legal and social location of women's medicine.

What were the minimum presuppositions that made this conflict about appropriate medical care possible? From our side, our theological beliefs and hopes cashed out in a single criterion for diagnostic action which seemed to us common-sense, but turned out to be increasingly hard to maintain in the contemporary medical context. We asked simply and directly that all treatment of mother and child be correlated to the medical interests of both. We did not consider termination a 'treatment' for our child, nor diagnostic testing which was not directed at a proximate and remediable medical problem. In theory, we would have been open to a discussion of termination on grounds of the safety of the mother, but this consideration was never relevant. Thus, in this first case of conflict with the medical profession over the appropriate use of medical testing, the conflict of interests was sustained from the medical side by the perceived necessity of establishing an accurate date of conception. This places the question in the foreground of whether there are serious medical reasons for establishing gestational age, or whether this is inherently tied to the pressure of legal deadlines for abortion.

Not satisfied with her experience at the *Frauenklinik*, Stephanie contacted a midwife to discuss having a home birth. The midwife conducted her visits in the simple ways of a previous generation: measuring growth, asking how Stephanie was feeling, looking at her colour, asking about specifics like aches and pains in the teeth and bones. The 20-week ultrasound scan also came up totally normal. Stephanie being in the pink of health both before and during her pregnancy, as well as still relatively young (29 years old), a home birth was allowed with little comment. To the day of birth, no countersigns indicated that anything but a normal baby was soon to arrive.

Adam was born one stormy August morning, three weeks early but apparently healthy. He nursed as expected despite his small size. Mother, father and midwife greeted the dawn with relief at the short night of labour and the arrival of a son sporting a complete complement of fingers, toes, eyes and ears. We remained together at home for the next three days adjusting to the rhythmless biological cycles of our new family member, interrupted only by occasional midwife check-ups.

On the fourth day Adam was scheduled for his first medical check-up, for which we were glad, as he seemed to have an infected navel and a bit of a rash. The doctor looked him over that midday and told us, 'New babies often have skin rashes. They can be completely normal and it will most likely go away.' An important transformation was soon to take place which should be noted here. Up to this point, despite his premature birth, Adam had been considered normal and so was only looked at with the cursory attention one pays to the normal ups and downs of infant life. He was, however, soon to become 'handicapped'; his every symptom a portent, and his 'future problems' not only mentioned, but dominating every examination and discussion of treatment. But on this fourth day at noon, he was still a normal child, with an apparently normal rash.

At seven that night, Adam went from being a bit listless and blotchy to grey, limp and lifeless as Brian held him and talked on the phone. Though reviving within a minute, he was, at Stephanie's insistence and against the midwife's intuition (for whom he was as yet a normal child who had probably just choked on some milk), soon taken to the local children's hospital emergency room for examination. It was Stephanie's insistence on medical testing which overcame the opinion of multiple medical personnel repeated at several stages of the process: 'If it will make *you* [i.e. the mother] feel better, we will check him over'. Having taken Adam in just for a check-up, it soon became apparent that this was no first-time-parent panic: Adam was in critical condition despite his relatively unworrying visible appearance. His blood oxygen was dangerously low and without attention he almost certainly would not have survived the night.

By midnight he was in intensive care with what turned out to be a serious blood infection, and the doctors told us that they could not with any certainty predict the outcome. Adam's odds of survival were simply unknown. Forty-eight hours later, and to our unimaginable relief, Adam had turned the critical corner. It was during the period of his ascent back from his drugged and betubed existence in his tiny box that Stephanie first noticed in his chart the words, 'verdacht Trisomy 21' – 'suspected Down's syndrome'. After asking the doctors if this meant what we thought it did (and trying unsuccessfully to discern when they planned to tell us of their suspicion), we were left to absorb our dark discovery: Adam is not normal.

The emotional turmoil this discovery caused was immense, and not simply because of the unexpected nature of the news. In part we were simply reeling from the rapid swinging of our fortunes. One minute we were coming to grips with the possible death of a beautiful healthy boy, the next, the possible life of one disabled. The disorientation caused by an immediate shift from 'Where do you bury a child in a foreign country?' to 'I'll never do the things I love with my firstborn son' should not be underestimated, and we were understandably disoriented. Thus, it should not have surprised us, either at the time or in hindsight, to have contemplated the horrible and yet predictably human question: 'Maybe it would be better if he died'.[4]

On the tenth day of Adam's hospitalization, we had our first discussion with the head physician about Adam's suspected Down's, a conversation which soon came around to the following statement: 'We took blood on admission to establish whether Adam has Down's or not, but the test failed. It will take about ten days to get the results of the new test back.' We asked whether it was the law in Germany for genetic tests to be undertaken only with permission, and he admitted without explanation or apology that yes, parental permission is legally required.

We made our position clear. We were unhappy that they had taken blood and done a genetic test without our permission, and were now simply informing us that the test would be repeated: on those grounds alone we decided out of pure stubbornness that it was not in Adam's or our best interests simply to sit back and let them proceed without discussion with us. So we refused the test, a decision which, at least while we were in the hospital, we were very happy to have made.

We decided from that point on that we would insist on asking a relatively simple question when a diagnostic procedure was suggested: 'Will it aid Adam's treatment?' This turned out to be a revealing question. The only reasons offered for having the genetic test fell under the category of 'for future planning'. This included testing for known problems suffered by children with Down's such as problems with sight and hearing, intestinal troubles and thyroid disturbance.

However, we soon found that the real reason to test our son's chromosomes was to know what *kind* of Down's he has so that we would be informed about our chances of having healthy children in the future. Now the point of the discussion and the push for testing began to emerge from the murk of scattershot argumentation: you wouldn't want to have any more of 'these children'.

It is worth noting at this point that this is a position we often hear expressed. We have been forced to conclude that this is something like the standard Western medical position, offered with some insistence and in good faith even by well-meaning Christian medical professionals in our families. Medical professionals have objected to the suggestion of this chapter that simply *having* genetic testing has shifted medical behaviour toward the disabled. 'Testing is not a judgement on disability', they chorus. Stephanie's second pregnancy again raised questions about this claim. One of Adam's primary medical carers, a specialist in Down's, and with whom we are very happy, was outspokenly certain that we were making a mistake not to have detailed testing of Stephanie's second pregnancy, because 'it would put your mind at ease'. When Stephanie replied that she still was not interested, the response presumed that the screening would occur in any case because we would agree to the routine regimen. This assumption set the physician at ease, because, in her words, 'Your risk is only one in 100 now, and they'll do very detailed scans on you anyway'. The scans being proposed might have been less invasive and dangerous to the foetus than amniocentesis, but the suggestion that we wouldn't want another child like Adam, and the 'treatment' being offered the newest member of our family, certainly were not.

In the end, we were not persuaded that the tenuous links between the genetic test and any possible benefit to Adam's health warranted us giving permission for a re-test. From our perspective, and given the weakness or even counterproductivity of the reasons offered for having the test, it looked to us as if the insistence on the test was grounded in it being part of a routine battery applied to children suspected as disabled. Given this, and the fact that the initial test had been taken without our permission, we refused consent to the repeat genetic test. Adam's suspected Down's remained just that – suspected.

Again we can uncover the logical presuppositions sustaining this conflict about genetic testing. From our side we were again acting on the criterion of refusing testing that does not directly further Adam's current treatment. From the medical side it appears that two main considerations played the most significant roles. The first was that such testing is routine, yields interesting and potentially useful information, and might be of some use in thinking through possible medical issues. There might be no direct benefits to Adam, but it is dimly possible that there might be some in the future and we (the medical profession) like to have the information. Second, you (the parents) might want to think about the future, and at the root of what you will want to avoid in the future is a second child with Down's. Both presuppositions transparently aim at the augmentation of medical power, not necessarily at the expense of the present patient but certainly not to his benefit.

Now began a psychologically fascinating period in which we experienced something like the tortures of those who do undergo much more rigorous foetal testing. Like those receiving a first diagnosis of possible Down's *in utero*, we could not be 100 per cent sure yet whether he had it or not. For us this yielded a constant debate about symptoms. Maddeningly, Down's is a syndrome exhibiting a phenomenally wide range of symptoms which many people without Down's may also have: creases in the hands, premature birth, low tonus, small size. For six months we only intermittently enjoyed Adam as we alternately scrutinized his symptoms and then tried to push them out of our consciousness. Behind that unstoppable mental oscillation lay an insidious crossroads. In one direction lay the question 'Do I betray my normal child by suspecting him of defect?' and the other, 'Do I betray my disabled child by withholding some treatment which he might need at this early stage?' It was a question which determined many details of parental life, and one we could not yet resolve.

These and similar questions are surely also present in the mind of those told that their child *in utero* is suspected of having Down's – with the one extremely significant distinction that has become *the* moral boundary characterizing the age of legalized abortion: our child was outside the womb. Having passed the birth canal he was indisputably a citizen, a right-holder and a person in the fullest sense. We find it hard to imagine standing at this crossroads if his face had not been visible, and if we were surrounded by reasonable authorities, families and friends telling us, 'Don't risk it. Start over with a clean slate.' Under such circumstances one imagines how easily any moral scruples about abortion fade into insignificance with the path so helpfully smoothed by everyone involved.[5]

So too did the ambiguity of the 'suspected' have a clear effect on us. In previous ages the only way to diagnose a borderline case of the syndrome would have been to 'wait and see', while monitoring a child's eating, sleeping and development. Their fears were tied to the high child mortality of the age, but this was a very different sort of fear from that of modern parents. Today the landscape of parenthood is radically different in that simply feeding and caring for an infant who is apparently growing and developing is cast as dangling on the edge of a dangerous precipice. The advent of testing brings the future to bear in a new and forceful way. This all-too-present future unsettled us, a fear learned from Adam's doctors. In an ironic and very revealing way, Adam's being 'suspected' played a heuristic role helping us to see the aims, skills and interests of the doctors examining Adam and so to question whether their fears coincided with our proper parental concerns. We embarked on a round of doctor visits to find someone to look after him, and found that he unnerved doctors increasingly unfamiliar with Down's infants and who instinctively reached for genetic testing as a diagnostic panacea.

The combination could and did become explosive on occasion. When Adam was five months old Stephanie took him to the doctor for a cold, fully aware that Down's children tend to respiratory-tract infections. The doctor looked at his medical records, and then looked at Adam. Then came a question, uttered in tones of moral outrage: 'Why haven't you had the genetic test done?' The result: a heated and deadlocked exchange of views, in which the doctor made bold to win Stephanie to her position, at one point even saying, 'You know, don't you, that Adam will never go to normal school?' The delivery of this verdict was the only result of the appointment: there was no further examination, no suggestion of treatment. Somehow, even from quarters where you might not expect it (this time a female homeopathic physician) the demands of the ideology of genetic testing managed to crowd out attention to the symptoms of infection, with the result that Adam was denied medical attention to his actual sickness. It turned out to be a cold, which Stephanie was left to deal with, alone, shaken and denied medical advice.

Here again our parental priority was that Adam be treated for the presenting sickness, on this occasion his cold. From the medical side the conflict could only have been sustained on the basis of a belief, or a fear, that treatment was not advisable without a proper, scientific (read genetic) conformation of diagnosis. This, in turn, yielded a medical perception that a child's parents were resisting a proper social understanding of the child, and so irrationally avoiding the genetic test that would both dispel their illusions of health and concretely document that the child belonged in the social and medical role assigned to children with this syndrome. The actual conflict ended with the stymieing of both objectives: Stephanie did not accept her social role, and Adam was not treated for his cold. Here the presence of the genetic test was the *condition* for the conflict between medicine as diagnosing a genome and medicine as treating a patient.

Through the recommendation of a friend in our church, we finally made a connection with someone who was both a neonatologist and the father of a child with Down's. He looked at Adam and said that, for several reasons he patiently explained to us, he felt there to be something like an 85 per cent chance that Adam had Down's. For the first time we felt as if someone with medical training had actually *looked* at *Adam*. Here was one of a dwindling number of medical practitioners with first-hand experience of the increasingly rare condition called Down's Syndrome, and who could therefore assess Adam with the eyes of experience. In addition, both he and his wife could give us the kind of parental advice we sought about how to deal with Adam's particular health difficulties. For the first time our crossroad dilemma seemed to be resolving. It was looking more likely that he did have Down's than that he did not, and more importantly, that we could handle it whatever the answer. We passed the news on to our families across the Atlantic who also, never having met Adam, would have tentatively to begin their own journeys of acceptance.

Adam seemed to be developing well, if slowly, and whether he did have Down's syndrome or not he showed only mild versions of the classic symptoms. We had been told during his first hospital stay that he had a characteristic hole in his heart, but from our limited viewpoint, and from what observers, including medical professionals, cared to express to us of any worries they might have had, the leakage caused by the hole seemed not to be particularly threatening. Having therefore put the possibility of surgery out of our minds, and having waited to have the cardiac ultrasound in order to see if the hole closed on its own, we were unpleasantly surprised when the scan results arrived: the holes between Adam's atria and ventricles were developing into a lethal combination, and he would immediately have to undergo open-heart surgery. The doctors were unanimous in urging surgery as quickly as possible to avoid irreparable damage to other organs due to massive imbalances in his intersystemic blood pressures.

By this time we had decided that if Adam would already be having routine blood work as part of his surgery, we would also allow the chromosome test. It turned out that the main factor in our decision was not a new conviction that Adam would receive better treatment as a result. At this stage the doctors, with our express permission, were simply treating him as having Down's, and had quit asking us to do the genetic test. When Adam went in to have his heart operation we reversed our position, not on medical grounds, but as a concession to two social pressures.

The first resulted from an aspect endemic to the maternal tribe, the desire to belong and to be able to hand on 'trade secrets' about child-rearing. Not knowing the exact state of Adam's condition meant that from the beginning Stephanie struggled with answering the basic questions that are inevitably presented to new mothers: 'How old is he?', 'What is he doing now?', and so on. As Adam's growth and development were slower than that of 'normal children', Stephanie felt she was constantly having to answer such questions without herself really knowing what to answer. The crucial point, however, came shortly before Adam's operation when another mother, after looking at Adam, said to Stephanie, 'My son has Down's Syndrome too'. This question crystallized Stephanie's growing need to leave this limbo and know to which community Adam, and herself as his mother, would ultimately belong.

The 'hard fact' of a genetic diagnosis would constitute a socially acceptable permission to begin to seek out other parents of children with Down's and begin to throw ourselves into learning about the condition. It would resolve our crisis of belonging. We felt caught between a medical establishment unwilling to say Adam had Down's without a genetic test, and the puzzled silence he continually evoked from most people, both inside and outside the Church. After the test, we soon learned, Adam would come to be admired for the opposite reason, as a fine example of that unfortunate race called Down's children.

The second reason for having Adam genetically tested was more pragmatic. In Germany social benefits are offered to parents of children medically certified as handicapped, and we were told that a genetic test would be required to secure Adam's diagnosis. An irony of this aspect of our experience is that now that we are in Scotland, we are discovering that certain forms of tax benefit are not readily extended to families with Down's children because they are not considered 'severely' handicapped.[6]

It was not, in the end, medical pressure which led us to permit a test which we felt was not a direct benefit to Adam's health. It was a sense of social isolation paired with a context in which the only person who dared to state the obvious based her knowledge on her own child and not a test. So our decision to test rested on the much messier and less defensible sense that the test would provide social benefits primarily to Adam's harried parents. We justified the decision on the grounds that within the context of another hospitalization, it would not cause Adam any new suffering, and would resolve social conflicts for *us*. Those conflicts were financial, emotional and interpersonal in that we wanted to be full participants in the social support system. What is clear is that neither for us nor for those around us did Adam straightforwardly belong: he certainly belonged as a child in all groups, but he was in limbo without the test in the subgroupings which provide the concrete support and education of parents.

Even we were surprised at the relief we felt at receiving the genetic diagnosis. The pressures of six months of either/or, of the crossroads of two looming alternate futures had weighed on us more than we could have articulated at the time. During this period we found that the social pressures for a 'definitive' diagnosis were stronger and more varied than we would ever have imagined. In the end, acceptance of Adam as he is was made easier for us by the fact that the major milestones in our learning of his condition coincided with his life-or-death moments. As Michael Berube puts it in his book *Life as We Know It*, in the wake of a life-and-death crisis, hearing that a living child has Down's 'seemed like a reprieve'.[7] Whatever challenges Adam's life would present in the future paled into insignificance beside the altogether more threatening possibility of his not existing at all. All life takes on a remarkable luminance against the backdrop of death.

A concluding unscientific postscript.

Having told our story, we would not presume to extract a moral for general application. The following observations are a loose collection of realizations which have been pressed on us by these experiences. The most overwhelming sense of continuity we found in these experiences of the impact of genetic testing, what might for us be called the 'soul' of the age, is best characterized as a pervasive and very particular form of fearfulness. Perhaps our most vivid impression is of the mismatch between the great anxiety about testing Adam when compared with the meagreness of the medical problems that it could be expected to resolve. Noting this disproportion allowed us to see the way our refusal to test was perceived as a threat to medical authority which assumed that the anxiety of medical practitioners would and should constitute a reason for us to be anxious and thus to comply with the recommended course of action.

The anxiety in question seems to have been generated by a range of possibilities: that we could not be trusted to do the 'best' for our child, that physicians were being denied the information they felt they needed to do their job, and the fear that medical authority was being displaced or threatened by our asking for medical interventions to be justified by criterion other than their own. In short, our behaviour constituted a challenge to the techniques which reassure medical personnel by providing a sense of mastery or control. The various medical professionals with whom we had contact must not be singled out as bad eggs on this account. On the contrary, it is their very best and conscientious humaneness which fuelled the conflicts we have narrated. They were expressing inherited versions of responsible parenthood which differed from ours. Their belief in the neutrality of genetic information masked a sharply normative claim that no *good* parent would resist genetic testing.

Some have protested that framing the story in this way betrays an unwarranted suspicion of medicine. This objection tends to overlook that we kept *returning* to medical practitioners for advice about our son, and that we thought this to be an important part of our parental responsibility. We are grateful and consider medicine a form of God's grace in having saved Adam's life not once, but twice, and through the most intensive and invasive techniques medicine has developed. Yet, despite the pervasive dogma that medicine is non-coercive, in our experience, *any* questioning or refusal of proposed courses of treatment were cause to place *us* under suspicion as recalcitrant, unreasonable, deniers of facts, etc. Irrationality seems most readily applied when the only medical reason for a test is that it is routine. Take the example of HIV testing during pregnancy. The transmission factors for the disease are well known. If faced with a pregnant medical professional (Stephanie being an example), one would expect their word to suffice that an HIV test is unnecessary. Yet the test is routine because medicine systematically distrusts people's statements about their health status. There may be good reasons for this distrust. But there are also good reasons to note how this corrodes the patient's belief that their words are being taken seriously. The impression is left that routine has displaced attention to individuals, and become the rationality against which only the irrational could raise questions, and only the suspicious would think to.

In the face of this atmosphere of anxiety, we have found Luther's meditations on Psalm 127 provocative and comforting.

> Unless the LORD builds the house, those who build it labour in vain. Unless the LORD guards the city, the guard keeps watch in vain. It is in vain that you rise up early and go late to rest, eating the bread of anxious toil; for he gives sleep to his beloved. Sons are indeed a heritage from the LORD, the fruit of the womb a reward.

Luther sees in these verses a contrast between human activity which takes place in trust, faith and appreciation of the variability and fecundity of God's love through creation, with human activity framed by anxiety about our efforts to create goodness. 'Because God gives him nothing unless he works, it may seem as if it is his labour which sustains him; just as the little birds neither sow nor reap, but they would certainly die of hunger if they did not fly about to seek their food.'[8] The same, Luther says, is true of human procreation.

This suggests that one way to begin to think about the dynamics of our experience of genetic testing is as a symptom of modernity's near-complete loss of the sense that human action is only *discovering* our sustenance and continuation as a species, nation or family, not a *grasping* or *creating* it. To think of ourselves in this way is to think of medicine as an activity of carers for what we have received. If we think we create, or are absolutely indispensable in the preservation of human life, then we become anxious with the anxiousness that marks those without faith.[9] To put things this way is to throw Christian views of genetic testing into a very different light, suggesting very different aims, and a very different attitude.

Luther rightly observes, long before Hannah Arendt,[10] that children are the wellspring of all society, and so we should expect the bearing of children to be a region of human activity which is especially prone to the anxieties of fear, and the lust for power of human Promethean urges. Procreation is either humanity's most important resource to be technologically 'managed' at all cost, a site of ultimate power struggle and angst, or it is a particularly rich seam of God's gift to humanity, the 'jewels in the mines': 'Like arrows in the hand of a warrior; so are the children of youth'.[11] Hauerwas is thus right to remind us that talk about the social forces which come to bear on the disabled and those who care for them is a way to address questions of social justice that circumvents the liberal, democratic definitions of justice within which it is hard to conceive of the handicapped as one of us.[12]

If Luther suggests that the beginning of a proper relation to the handicapped is a sense of the gift-nature of all human life, Augustine provides a bit more detail about how to understand the 'otherness' of the handicapped as a divine gift to humanity. In the midst of a discussion about the limits of the human race, Augustine stresses that being human is not defined by criterion of physical or mental perfection which 'deformity' vitiates, but is grounded in descent from Adam.[13] He therefore concludes that, far from approaching the 'different' as defective members of the race, and to be shunned, we should instead see them as God's special works, because God 'knows how to weave the beauty of the whole out of the similarity and diversity of its parts. The man who cannot view the whole is offended by what he takes to be the deformity of a part; but this is because he does not know how it is adapted or related to the whole.'[14] Later he suggests that this creaturely otherness is a sign of God's otherness, and a proof that we should expect the unexpected from him, supremely, resurrection from the dead.[15] This eschatological openness to the disabled is well encapsulated in Karl Barth's comment that 'the value of this kind of life is God's secret. Those around and society as a whole may not find anything in it, but this does not mean that they as a society have a right to reject and liquidate it.'[16]

For Augustine then, human otherness and difference in all its variety is part of God's way of keeping us from falling into the wonderless idolatry of homogeneity. Like Luther, Augustine sees the fault-line running between those who see diversity as a gift, and those for whom it evokes anxiety. Perhaps the polarity gives us a way to understand Bonhoeffer's comment, written amidst the anxieties and violence engendered by the quest for a perfect race, that the strong are not made stronger by euthanizing the weak. The strong have been made stronger by understanding the value of those consigned to the outside by our merely human visions of perfection and human community.[17]

Though not putting it in these terms, perhaps what Bonhoeffer was looking for was a way to say that living with the handicapped is beyond our means. Those who live with them know this. Statistics tell us that most parents abort handicapped infants because they feel they don't have the resources to raise them.[18] But those who embrace their children (any children) as having been *given* defy their urges to protect themselves against them and so open themselves to the Spirit's work. This is not to coerce the Spirit's help by asking for it and acting as if we will receive it. The Spirit's most surprising and essential work is to comfort us by transforming our ideas about who we are and what we can do. We learn that the abilities we need to flourish are abilities we don't *have* but must be *given*, and therefore must *ask* for. Theology needs to abandon its false modesty about the resources promised in the Spirit. When it does not, it cannot avoid the trap of wishing for perfection from humanity. Either we say that the Father gives through the Son and helps us to receive in the Spirit, or we end up hoping with the world for the least complicated experience of parenthood, marriage, work, and so on. The problem of not being able to receive the handicapped with open arms turns out not to be an insufficiently inclusive anthropology but an atrophied pneumatology.

Perhaps if Christians inhabited this understanding of disability they might, for instance, buck current ways of narrating life with handicapped children. Bookshops are filled with stories in which parents-to-be conceive with the aim of producing a perfect child, and so narrate their story with their handicapped child in terms of how difficult it was to adjust to the letdown of disability.[19] How different the story could be if told within a culture that took children, all children, and especially the different, as gifts. Then we might hear stories such as that suggested by Luther, Augustine and Bonhoeffer, in which the perception of someone as deformed gives way to a new realization that deformity is in the eyes of the beholder.[20] Such an approach does not minimize the importance of medical devotion to ameliorating all peoples' physical problems. But it is to urge doing so by seeing clearly that either our faith and gratitude or our fears will shape the political strategies we pursue. In relation to the genetic test, our experience suggests that, for a range of possible reasons, contemporary medicine finds itself in a framework of fear rather than gratitude, and so finds it difficult to separate medicine as a project responsible for creating 'normal' children from medicine as a human technique for caring for each person's physical problems.

Notes

1. George Grant, *Time as History* (Toronto: University of Toronto Press: Anansi, 1995), p. 14.
2. George Grant, *Technology and Empire* (Concord, Ontario: Anansi, 1969), p. 137.
3. This is a recurring concern, most fully expressed in 'Timeful Friends: Living with the Handicapped', in *Sanctify Them in the Truth: Holiness Exemplified* (Nashville, TN: Abingdon Press, 1999), pp. 143–56. Also John Swinton (ed.), *Critical Reflections on Stanley Hauerwas' Theology of Disability: Disabling Society, Enabling Theology* (Binghamton, NY: Hayworth Pastoral Press, 2004), pp. 11–25.
4. Søren Kierkegaard rightly suggests that this is indeed an all-too-human

thought, and just so: sin. If Abraham had not loved his son Isaac, he would have been a murderer like Cain. Søren Kierkegaard, *Fear and Trembling, Repetition* (trans. and ed. Howard Hong and Edna Hong; Princeton, NJ: Princeton University Press, 1983), p. 74. Kierkegaard's point is to draw out how easily subvertible supposedly disinterested parental love actually is, suffering as all love does under the 'aesthetic illusion of magnanimity' (*Fear and Trembling*, p. 93). See Amy Laura Hall, *Kierkegaard and the Treachery of Love* (Cambridge: Cambridge University Press, 2002), p. 71.

5. Barbara Katz Rothman's book, *The Tentative Pregnancy: Amniocentesis and the Sexual Politics of Motherhood* (London: Pandora, rev. edn, 1994), is an important contemporary study of these forces, and a stinging indictment of the claim that prenatal testing facilitates parental bonding with their unborn children. Rather,

> Increased knowledge, without increased responsibility on the part of the society, translates to increased knowledge with inevitable responsibility on the part of mothers. We are asking mothers to become the gatekeepers of life. We are individualizing social problems of disease and disability, medicalizing life itself, and doing it on the bellies of pregnant women (p. xiv).

6. Department for Work and Pensions, Disability and Carers' Service, Disability Living Allowance.
7. Michael Berube, *Life as We Know It: A Father, a Family, and an Exceptional Child* (New York: Pantheon, 1996), p. 5.
8. Martin Luther, 'Exposition of Psalm 127, for the Christians at Riga in Livonia', in Walther Brand (ed.), *Luther's Works, American Edition*. VL. *Christian in Society II* (Philadelphia, PA: Fortress, 1962), p. 326. The insight is based on many biblical texts. Augustine makes the point by quoting 1 Cor. 3.7 in *The City of God*, 22.24.
9. Luther makes the point elsewhere in a comment on the prayer of thankfulness for healing (through medical treatment) of Hezekiah in Isa. 38.10-20:

> for you heard in the song above that all things are kept and cared for by the Word. Then the ungodly cry: 'If the word does everything and provides nourishment for everything, we do not want to eat or take medicine.' For them he takes up this example. As for you, make use of means. Do not rely on them but use them, since God has created them. If they do not help, commit the matter to God. Do not say: 'Doctor, if this will not help this time, I refuse to take it anymore.' Yes, you want to have your own way! So we all go beyond the proper use of means by clinging to them . . . Others despise works altogether, so does this song ascribe the power to the Word and not to the medicine. Yet it does not forbid that we use them, but the prophet's example supports their use, since he poultices the wound with a cake of figs. Martin Luther, *Luther's*

Works, American Edition. XVI. Lectures on Isaiah, Chapters 1-39 (ed. Jaroslav Pelikan; St Louis, MO: Concordia, 1969), p. 344.
10. Hannah Arendt, *The Human Condition* (Chicago, IL: University of Chicago Press, 1958), ch. 6.
11. Job 28.1–6; Ps. 127.4.
12. Stanley Hauerwas, *Performing the Faith: Bonhoeffer and the Practice of Nonviolence* (Grand Rapids, MI: Brazos Press, 2004), p. 230.
13. Here anticipating Robert Spaemann's similar point. See Berndt Wannenwetsch's contribution to this volume (Ch. 11).
14. Augustine, *City of God*, 16.8.
15. Augustine, *City of God*, 21.8.
16. Karl Barth, *Church Dogmatics* 3.1, pp. 423–24.
17. Dietrich Bonhoeffer, *Ethics*, in Ilse Tödt, Heinz Eduard Tödt, Ernst Feil, and Clifford Green (eds), Reinhard Krauss, Charles West, and Douglas Scott (trans.), *Dietrich Bonhoeffer Works* (Minneapolis, MN: Fortress Press, 2005), VI, p. 194.
18. See Amy Laura Hall's contribution to this volume (Ch. 5).
19. Danny Mardell, in Sally Weal (ed.), *Danny's Challenge: The True Story of a Father Learning to Love his Son* (London: Short Books, 2005), and Mitchell Zuckoff, *Choosing Naia: A Family's Journey* (Boston, MA: Beacon Press, 2002) are sobering examples of such a parental narrative which, at different levels of intellectualization, both portray the struggle to justify and respond to parenting a 'different' child conceived in many ways as a burden.
20. In this, Barbara Kingsolver's fictional commentary is right on the mark:

> Mama Mwanza almost got burnt plumb to death when it happened but then she got better. Mama says that was the poor woman's bad luck, because now she has got to go right on tending after her husband and her seven or eight children. They don't care one bit about her not having any legs to speak of. To them she's just their mama and where's dinner? To all the other Congo people, too. Why, they just don't let on, like she was a regular person. Nobody bats their eye when she scoots by on her hands and goes on down to her field or the river to wash clothes with the other ladies that work down there everyday.

Later, she articulates the reflections of a disabled Westerner who craves such acceptance.

> The arrogance of the able-bodied is staggering. Yes, maybe we'd like to be able to get places quickly, and carry things in both hands, but only because we have to keep up with the rest of you, or get the [punishment]. We would rather be just like *us* and have that be all right. (Barbara Kingsolver, *The Poisonwood Bible* [London: Faber & Faber, 1999], pp. 60, 559)

2

'What's Wrong with You?' Disability and Genes as Ethics

Christopher Newell

The moral encoding found in the everyday

'What's wrong with you?' It's amazing how people I don't even know will quite confidently sidle up to me, look at my obvious sign of social status – disability – and with a caring tone of voice ask me those famous words, 'What's wrong with you?' Radio and TV presenters see it as such an important question. If I reply – as I am given to – with something similar to, 'I'm fine thanks, and how's your sex life?' the demeanour and situation changes markedly. At first thought an obscene dissonance, yet is there really anything very different in terms of the nature of the enquiry? The knowledge we as a society have about disability ('What's wrong with you?') has largely agreed-upon social dimensions. In such a small statement we encapsulate so much about our notions of health, disability, the good life and especially normality. We also encounter the centrality of the disabled body to the narration of ourselves, humanness and our notions of what is nice, normal and natural. This is a paradox where disability is at the same time highly central and yet marginal at best. The centrality of disability may be found in the way in which it features in headlines about medical cures for disability and in justifications for charities and welfare businesses. Yet it is marginal in that rarely do the many issues and voices of people with disability feature.[1]

In moral terms my disabled body, and its status, is public property. Of course to some extent this is understandable. I am one of the few people who identifies as living with disability who also teaches and consults in bioethics. I have obvious signs of disability, such as using a wheelchair for mobility and periodically needing oxygen and other therapy. In Australian society, it is rare not only to have an ethicist with disability but even more an academic and professional with disability.[2] My work in bioethics and health care in general emerges from many years of experiencing institutional care as a patient, and a profound desire to tackle the power relations and practices which disable.

While I obviously see my life situation as private, this stands in stark contrast to the very public dimensions found in the use of a wheelchair and other obvious signals referred to above. Such a situation is in interesting contrast with the social and normative dimensions of asking about a person's sexuality – something intensely private among those who fit within the norms of sexuality and yet something explicitly for the tabloids if we depart in some way from the norm. Disability is crucial to the narration of the norm. It provides the antithesis to normality, something so self-evident we rarely have to think about it. For example in the expressions I often hear such as 'Thank God I do not have a disability'. Indeed, as Lennard Davis suggests in his rethinking of normality, 'the problem is not the person with disabilities; the problem is the way that normalcy is constructed to create the "problem" of the disabled person'.

There is a sense in which such a taken-for-granted phrase as 'What's wrong with you?' encapsulates many of the things that I would suggest are wrong with society, bioethics, theology and the way in which we narrate disability and genetics. On a wider scale, the Disability Rights Movement has been deeply disturbed at the encoding found in such taken-for-granted approaches to disability and its construction within accounts of theology, bioethics and indeed medicine as a project. Accordingly, in this chapter I argue for an account of a common humanity and approach to ethics and public policy which starts with the question 'What's right with you?' Such an approach has important implications for our common future. In order to make my point clear, allow me to tell some more stories.

'Community health and medicine 105'

In teaching the above course in undergraduate medical ethics, one of the first things that I do with my first-year students is to help them explore the troubling and contested world of disability by critically examining the work of Joseph Fletcher in his stimulating book *Humanhood: Essays in Biomedical Ethics*. One of the early developers of bioethics as a discipline, Fletcher was at the time an Episcopalian priest. Although he was later to revise his stark claims in this work, Fletcher identifies what he terms 'fifteen positive propositions and five negative propositions' relating to what he refers to as 'a profile of man' (somehow women tend to miss out). His fifteen positive propositions can be summarized as

1. minimum in intelligence (i.e. an IQ of 40)
2. self-awareness
3. self control
4. a sense of time
5. a sense of futurity: 'Sub human animals do not look forward in time: they live only on what we might call the visceral strivings, appetites'. He cites William Temple as emphasizing 'purposiveness as a key to humanness'.
6. a sense of the past (i.e. memory)
7. the capability to relate to others
8. concern for others

9. communication
10. control of existence
11. curiosity
12. change and changeability
13. balance of rationality and feeling
14. idiosyncrasy ('to be a person is to have an identity, to be recognizable and callable by name')
15. neocortical function.[3]

In other words, it may be seen that Fletcher's account revolves around a person being in control, rather than with being embodied.

As part of my first-year lectures I routinely select a volunteer, or at least someone sitting in the front row! One by one I remove from this student human attributes that Fletcher describes as been essential for claims to being a person. As each aspect is removed I ask the class whether or not the necessary attributes for claiming personhood have been removed. It prompts discussion about whether there is a difference between human attributes and being a person.

As each of the criteria are removed, we agree that these are important aspects of humanness – aspects we as a society value very highly in the narration of what it is to be a person. This is evident in media accounts of the loss of attributes such as hearing, vision and memory, with those who have these losses being narrated in terms of overwhelming deficit. Such attributes are crucial to many taken-for-granted valued activities by non-disabled people. Yet in exploring this with the students it is very difficult to say that any one of them immediately removes our claims to being a person. Most importantly of course, the unwitting student immediately removed to the status of unperson loses the ability to communicate his or her (or should I say *its*?) thoughts. In the name of ethics we are deciding who belongs to 'us' as human beings and what constitutes the 'other'.[4]

Two accounts of personhood

At their best, accounts of theology and ethics offer us a holistic picture of the world and personhood. Yet Fletcher as a theologian and early bioethicist helped to contribute to the rise of accounts of bioethics that uncritically use dominant models of reductionist[5] thought, influencing ethical accounts of biotechnology to this day, particularly in terms of genetics. Paradoxically, under the name of *agape*, Fletcher contributed to the rise of utilitarianism as a dominant paradigm and a world which is so good at constructing those with a variety of impairments as 'other' as opposed to embracing those who do not fit narrow norms as part of 'us'.

In a striking contrast between worldviews, it is worth noting that about the same time Fletcher was writing his work, another American theologian, Paul Ramsay, was also writing, eventually contributing his enormously important *The Patient as Person* to the rise of bioethics.[6] It is an interesting commentary that Fletcher's reductionist utilitarian account, which operates within and perpetuates the power relations of modern biomedicine, is still part of mainstream thought, and yet Ramsey's questioning of this is little more than a little footnote in a secular bioethics text. As I use the concept of power, I should explain I am here indebted to the works of Foucault, who, following Machiavelli, explores in a variety of his works the complex strategic dimension of power in a given social setting. As it is deeply structural, it involves both constraint and enablement.

Contested knowledge

When we consider notions of genetics, healthcare, disability and humanness, we need to recognize that these are contested sociopolitical spaces. They arise as a response to historical processes and are discursively shaped.[7] We need to understand the area of genetics as a consistent development from the days of the Enlightenment and associated mechanistic and rationalistic approaches to the definition of the body. We think of the body fitting and working together as one complete machine in ways similar to mechanistic cogs, composed of larger and smaller parts which mesh together, *designed* in an objective form of knowledge we know of as science.[8] Hence, contrary to the media hype, I want to suggest that genetics is currently contributing to, and being used as a form of, creating otherness, and that disability is perhaps the exemplar of that otherness.

Interestingly, in my experience, the most traumatic thing that you can do to a medical student is not actually the exercise I have discussed before. Rather, it is to present a case to a student where there are absolutely no signs of pathology whatsoever. In the medical school in which I teach, medical staff are so focused on the teaching about pathology and on the importance of the reductionist account of pathology that a case where there is no pathology apparent is an anomaly. We utilize such cases in teaching since of course people present to general practitioners not only for treatment of disease but also for a check-up or insurance examination. Of course, the refreshing thing about genetics as a project is the discovery that we all have genetic variation and the potential for our genes to express in particular circumstances as disease.

Stories and genetic knowledge

There are several purposes for my telling such stories. In the first place, we encounter the multidimensional aspects of life from different perspectives, as well as start to explore the enduring metanarratives which shape the projects that are biomedicine, bioethics and theology. The noted narrative theorist Hilde Lindemann Nelson helps me to understand that when I challenge dominant accounts found in 'What's wrong with you?', I encounter a major problem of the power relations of knowledge:

How freely we can exercise our moral agency is contingent on a number of things. Most broadly, it depends on the form of life we inhabit: the niche we occupy in our particular society; the practices and institutions within the society that set the possibilities for the courses of action that are open to us; the material, cultural, and imaginative resources at our disposal; the constraints arising from the moral flaws within our roles and relationships; the shared moral understandings that render our actions intelligible to those around us. More specifically, the extent to which our moral agency is free or constrained is determined by our own – and others' – conception of who we are.[9]

In recounting my traumatizing actions upon my students of slowly removing the various human attributes important to the narration of personhood, we encounter a similar problem to that associated with reductionism and genetic thought. An early commentator, Rosalyn Diprose, points to the way in which genetic information has been constructed by geneticists within their paradigm[10] despite uncertainties within genetics. In her stimulating critique of dominant genetic thinking she argues:

> The search for the origin of identity and difference, therefore, cannot stop with the gene – an entity which the search itself produces. We are referred beyond genes to mysterious 'regulatory mechanisms' which oversee their production and spatio-temporal distribution. Nor will the origin be found there. As each origin dissolves into its other, we get closer to where we began: to the manifestation of differences at the surface of the body and between bodies, to their socio-political distribution and to the author of that distribution. However, we cannot find an ultimate author or subject of this system of differences either. The geneticist, like the gene, is a place holder, being also an effect of the same spacing and is, therefore, constituted only in being divided from itself. The body may be the self expressed, as Schenck would put it, but only with a lack of certainty – only by being inscribed in a system of differences which genetics helps to produce and maintain. In other words, we find that both the 'subject' and the 'object' of genetics, the 'world' and one's 'being' are constituted in relation to each other and are, therefore, always other than themselves.[11]

Likewise, Hubbard and Wald in their early book *Exploding the Gene Myth* pointed out years ago that the knowledge associated with the prediction of disease states such as cystic fibrosis has proved to be complicated, possibly resulting from different mutations in the DNA sequence. One is reminded of the age-old search for a technological fix and of social Darwinist tendencies in searching for inherited tendencies based upon our notions of what is undesirable. We see this even now with regard to the search for alcoholism and drug-dependency in genes. As Hubbard and Wald wrote years ago but in ways relevant to today:

the faith that genes for all sorts of troubles and conditions can be identified and isolated, coupled with the hope that this will lead to profitable diagnostic tests, is likely to continue to fuel the search for relative bits of DNA. Not only will this not cure or prevent the conditions, it will create a new group of stigmatized people, the 'asymptomatic' or 'healthy unwell', who, although they have no symptoms, are considered likely to have a particular disability at some point in the future.[12]

Though it has been some years since this critique was written, it certainly would appear that people are increasingly concerned at the implications of 'healthy unwell' found in discussions of discrimination based upon genetic knowledge. Countries around the Western world have uncovered and discussed practices that allow discrimination based upon the likelihood of genetic conditions developing. Parliamentary inquiries have condemned discrimination based on genetic knowledge as holding back 'progress' and have suggested legislative measures to protect people from discrimination. Yet, as someone who daily experiences discrimination as a person with disability, where legislation which supposedly delivers justice for people with disability is noteworthy for its exclusions, I am amazed at the failure to discuss existing discrimination and gaps in legislative protection. When we analyse the conversations we find some reinforcement of accounts of 'us' where the 'healthy unwell' are counted, and 'other' where more stereotypical bodies belong.

Some of us have long argued that a category of 'healthy unwell' has existed for some time, in that people who identify as members of the 'deaf community' (predominantly people who are born deaf and use sign language as their first language) see themselves as healthy and not disabled, yet the dominant worldview sees their situation as undesirable and needing medical intervention. Such an idea has recently been pursued fully by Davis (1997),[13] who, unusually for a non-disabled writer in the area, recognizes deaf culture as a form of understanding the world in her analysis. However, while she maps the issues, it is quite clear that she rejects the holistic notion, supported by a sociolinguistic analysis, of the deaf culture being a full culture. Likewise, she approaches the analysis of this issue quite clearly from the perspective of the hearing world and rejects the legitimacy of the claims of deaf culture.

Indeed, even geneticists recognize the complexities and hype that sometimes accompanies media coverage of the benefits of genetics. Under the title 'Will Genetics Revolutionize Medicine?' two geneticists, Holtzman and Marteau, argue in the *New England Journal of Medicine*:

> In our rush to fit medicine with the genetic mantle, we are losing sight of other possibilities for improving the public health. Differences in social structure, lifestyle, and environment account for much larger proportions of disease than genetic differences . . .
>
> Although we do not contend that the genetic mantle is as imperceptible as the emperor's new clothes were, it is not made of the silks and ermines

that some claim it to be. Those who make medical and science policies in the next decade would do well to see beyond the hype.[14]

Embrace or exclusion?

All of this points to the way in which knowledge is constructed, and how society has become less tolerant of a variety of forms of differences. As Mary Johnson, an American disability activist, has argued in a stimulating piece with regard to the abortion of so-called defective foetuses:

> This is not a discussion about a woman's right to choose. It is a discussion about the thinking that prompts the woman, or the couple, to make certain specific decisions based on cultural assumptions that have been shaped by discriminatory practices and attitudes – against disabled people . . . A decision to abort based on the fact that the child is going to have specific individual characteristics such as mental retardation, or in the case of cystic fibrosis, a build-up of mucus in the lungs, says that those characteristics take precedence over the living itself. That they are so important and so negative, that they overpower any positive qualities there might be in being alive.[15]

Johnson goes on to argue that society has an obligation to work to create a status quo which sees conditions such as cystic fibrosis as needing to be treated, not feared. In short, there is little tolerance of anything that deviates from a narrowly constructed norm. Parents and potential parents are faced with stereotypical information, and as Mary Johnson persuasively argues: 'A disabled fetus represents for parents a problem that may have far more to do with society than with disability. Disabled children confront a hostile environment.'[16]

Over the years I have seen the cautionary arguments of critiques of a biomedical account largely ignored. For some of us the ethical, legal, social implications (ELSI) components of the Human Genome Project really amounted to little more than an opportunity for a variety of legal players to travel the world at enormous cost justifying eugenic practices.

In exploring these dominant accounts of that which is nice, normal and natural that lead to everyday understandings found in 'What's wrong with you?', we need to see eugenics as an inevitable outcome of dominant discourses and power relations. This is certainly attested to not only in the rise of secularist utilitarian accounts but also in the everyday eugenic accounts that assail us in headlines, where eugenics is not named but where the values are found in the practices and approaches adopted. One example is found in articles which report the ability to screen for disabilities such as Down's syndrome. Some of my friends have been identified as having this condition, and the tacit assumption that it is better that such people did not exist reveals a lot about narrow norms. It can also be seen in the suggestion often made and rarely questioned that screening on the grounds of sex is an undesirable 'social' reason, yet screening on the grounds of disability is portrayed as an understandable 'medical' reason.

One of the most energetic writers for the reassessment of eugenics is Richard Lynn, who points out that in the first half of the twentieth century 'virtually all biological scientists and most social scientists all did eugenics, and so also did many of the informed public', but in the second half of that century support for eugenics declined. Yet his book argues for the reintroduction of eugenics and what he calls its eight core propositions. These may be summarized as follows:

1. The most important human qualities are health, intelligence and 'moral character'.
2. These human qualities provide the foundation for a nation's intellectual and cultural achievements; quality of life; and its economic, scientific and military strength.
3. Health, intelligence and moral character are substantially genetically determined, and it is possible and desirable to improve these qualities genetically.
4. Since the second half of the nineteenth century the populations of the Western democracies have been deteriorating genetically with respect to health, intelligence and moral character. This threatens the quality of civilization and culture and the economic, scientific and military strength of the nation-state.
5. It is feasible to improve the genetic quality of the population in terms of health, intelligence and moral character. 'Positive eugenics' will increase the numbers of children of the healthy, the intelligent and those with strong moral character; whereas 'negative eugenics' will reduce unhealthy, low intelligence and children of weak moral character.
6. The new eugenics uses human biotechnology to achieve eugenic objectives.
7. Eugenics serves individuals and nation-states. ('It serves the needs of the nation state because a nation state whose population has good health, high intelligence, and good moral character is stronger and more likely to succeed in competition with other nation states.')
8. The prohibition of biotechnologies in the Western democracies will not be successful. ('No new technologies that serve human needs have ever been successfully suppressed.')[17]

In his book Lynn also explores how the 'eugenics of human biotechnology' is likely to evolve, suggesting that it is 'likely to be used for the development of national strength by authoritarian states, leading ultimately to the establishment of a world state'.[18] Ultimately this is an issue incorporating politics and economics in the discursive shaping of public policy. These considerations will see whether or not eugenic elements are incorporated into public health 50 years from now. Given the developments identified, this certainly seems an insidious possibility. Of course, whether or not a world state emerges, the issue remains that people with genetic characteristics are being constructed as inferior, seen as a burden on the state and as impeding progress. Ultimately disability may be seen as the enemy of choice and potential, as Lynn constructs it. Yet he pays little attention to moral characteristics of people such as selfishness or selflessness, which will also impact upon the state.

Power and policy

In seeking to understand such challenges, we need to explore the way in which power is central to the shaping and enactment of policy, its implication in dominant accounts of theology and of bioethics, the rise of individualism in a post-Enlightenment world, and the way in which reductionist thought combines with economic and technological determinism. Our dominant accounts of medicine and biotechnology[19] help to reinforce the existence of good and bad genes, which is powerfully coupled with the rise of autonomy as the dominant value to be found in Western society. In all of this disability may be seen both as central, in justifying the claims of biomedicine and genetics, and yet marginal, as people with disability are talked about rather than listened to as experts. The one single exception to this is where our voices are utilized to magnify the tragedy of disability, exemplified by the Christopher Reeve approach to disability.[20]

Michael Cook responds to these sorts of developments, in particular the way in which our system of ethics review at the institutional level is increasingly being used to set social policy. In the following article, provocatively and accurately entitled 'Designer babies? Don't leave it to the bureaucrats to decide', he comments:

> I feel gobsmacked. It reminds me of the opening scene in The *Hitch-hiker's Guide to the Galaxy* when a Vogon spaceship announces that the planet is about to be annihilated. 'People of Earth, your attention, please. As you will no doubt be aware, plans for the redevelopment of personhood, family, and sexuality and morality require the building of a hyper-spatial express route through your values, and regrettably they have been scheduled for demolition. An independent ethics committee has given its approval. The process will take slightly less than two of your Earth minutes. Thank you.'[21]

While Cook is right to raise the vital issue of ethics committees being the *de facto* setters of social policy, I'm not certain that this is in anyway a recent phenomenon. Many of the developments in genetics and biotechnology in general rest upon the lives (portrayed as catastrophic) of people with disability. Yet, somewhere along the hyperspatial express route people forgot to ask those of us with disability about this – something that would be risible were it not of course entirely normal. Indeed, we have failed to explore how we discursively create disability via our accounts of science and ethics based upon underexplored accounts of normality.

In the last few hundred years, in losing our religion – or at least giving new shape to our spirituality and expressing that in different ways – we have increasingly lost any agreed-upon language with which to address these issues. As David Wells comments with regard to the changes which have occurred in Western culture since the Enlightenment:

> In the wider society, during the 18th and 19th century, the classical virtues came under fire from Enlightenment ideology, the Christian virtues in particular came under heavy bombardment, and slowly our language began to change . . . These classical virtues had always been thought about in relation to the community in which a person lived. To act justly was not an internal attitude but the practice of what was upright in a context where that moral virtue had been put to the test. When we come into the modern period, and as communities begin to disappear, the virtues come to stand alone, out of the social context in which they had formerly been understood. Thus, as MacIntyre points out, the virtue of honor increasingly comes to be understood in terms of a social status that is not awarded because of moral desert but gained through wealth or birth. When the virtues were thus privatized, when they were disengaged from public life, that life had to be governed, not by morality but by social rules that became etiquette. It was these rules that replaced the virtues, and these rules have now been replaced by governmental regulation and by litigation.[22]

Indeed, we have seen the rise of Thatcherite notions whereby there is of course no society, just individuals. We live in a world where the dominant – the only – value seems to be summed up by the cry 'autonomy rules, OK'. At the same time we are a death- and disability-denying society, one whose dominant approach to disability is found in these catch-cries:

Better off dead than disabled.
Thank God I don't have a disability.
Disability is about 'the Other' rather than 'US'.

Further, in addressing the issues of genetics we need to recognize our connectedness to the broader biotechnology futures. The recent embryonic stem-cell debate provides us with some important lessons.

The embryonic stem-cell debate

Goggin and Newell have elsewhere analysed the role of the media in perpetuating myths associated with embryonic stem cells causing superman to walk – and indeed fly – again tomorrow. Much of the heat has been generated with regard to the status of the embryo. Just like a variety of other biotechnologies, disability is at the heart of the claims of this new technology. As with other forms of biotechnology, there is a formula that seems to be played out in the media battles to secure this latest technology. It reads something like this:

1. The tragic life of an individual with disability or several devalued individuals is portrayed in a way designed to elicit maximum effect.
2. A technology is portrayed as delivering a person from disability, provided that society legalize, fund, or embrace such a solution.
3. Securing the technology means that disability has then been 'dealt with'; after deploying such rhetoric there is to be no more appeal to emotion, and the solution lies in the rational pursuit of the technology identified in step 2 (effectively there is only one, inexorable, logical step).
4. Disability as a political issue goes away, until next time it is needed in the powerful politics of media representation.[23]

In order to tackle the challenges posed by biomedicine in general, including genetics, we need an account of humanity which commences with the dignity of the human person, and then ask 'What stems from such inherent worth?' In this way, we also need to encounter the social contexts of disability, the limitations of genes, and reclaim an account of virtue which assists people to be all they can be. In so doing we will see that it is not so much a matter of putting a stopper back on the genie bottle, but of reclaiming disability as part of humanity and as a central starting-point for the projects which are bioethics and theology.

In so doing, it is worth moving beyond an account of disability as tragedy found in dominant approaches. As someone with a variety of impairments, I find that I am a better person and teacher because of my experience, even if, had I been asked, I would not have chosen such a road. Earlier this year I yet again had a few months in hospital, coming close to death. In so doing I had to reflect upon life and its purpose. For me, the very struggle with limitations and mortality is a major piece of learning. Sadly our focus on technical fixes often means we fail to ask about the importance of limitations and pain. Yes, I certainly took advantage of medical science to keep me alive. Yet, on reflection, I have come to realize how such limitations are part of a common human journey which has spiritual and physical dimensions. As journalist and disability activist John Hockenberry suggests in a delightfully countercultural way:

> Physical limits are a natural binding force in society, bringing people together. The arrogance of presuming that physical limits are somehow in opposition to life and to be hidden away is tragic. When people succeed 'despite their physical limitations', just as when crips 'have the courage to go on despite their disability', they are celebrated by the group. But when people's physical limits have become obvious, they expect to be shunned and left to their solitary self-hatred. It should be just the other way round. Separating oneself through personal triumph over some physical limitation is an act of isolation that repudiates the influences of family and community; openly acknowledging limitations binds and draws people together, as an emblem and reminder of just how similar we all are.[24]

Of course, such an account demands an examination of why we use medicine, science and genetics. It moves us beyond an automatic response to the disasters of life to find meaning in the very situations we know to be hostile to the human condition. It means encountering our ceaseless quest for normality found in genetics as ultimately futile.

Based upon such a premise of the inherent dignity of the human person, I would argue that we can then start the fragile and difficult conversations of what it is to be human and how we can use biomedicine to support and embrace human life. Such conversations demand an account of virtue in exploring the muckiness of life. Indeed, such an account of ethics, theology and science as involving respectful relationships has much to offer all people. Perhaps most importantly, such an approach helps us to encounter those we know to be 'other' as 'us', part of our moral community.

In identifying current practices as being lived ethics, we also have the opportunity for reclaiming ethics and theology as involving relationships, and via those relationships to foster transforming accounts of what it is to be human which moves beyond the largely taken-for-granted accounts we now know to be genetics and eugenics. Perhaps only then will we have an account of humanity, and everyday interactions, which starts with 'What's right with you?'

Notes

1. See G. Goggin and C. Newell, *Disability in Australia – Exposing a Social Apartheid* (Sydney: University of New South Wales Press, 2005).
2. While the UK preferred terminology is 'disabled people', recognizing that society disables people with impairments, in Australia we strongly prefer 'people with . . .' terminology, placing emphasis on being people first. The conversations have been similar, leading to different terminology.
3. After Joseph Fletcher, *Humanhood: Essays in Biomedical Ethics* (Buffalo, NY: Prometheus Books, 1979).
4. For further exploration of this see Christopher Newell, 'Retarded Children or Retarded Ethics?', in *Journal of Religion, Disability and Health*, 8.3-4 (2004), pp. 141–47.
5. Reductionism asserts that the nature of complex things is found as the sum or sums of simpler, or more fundamental, things. It is a key concept in medical science and the description of the body and its composition.
6. Paul Ramsay, *The Patient as Person* (London: Yale University Press, 1970).
7. The concept of 'discourse' is utilized in a variety of ways in the *social sciences*, constituting an institutionalized way of thinking, social boundary defining what can be said about a specific topic, or even a possible truth. Discourses affect and constitute our views on all things. After Foucault, I utilize *discourse* in a way which is closely linked to an account of power and state, since defining discourses involves defining reality itself. As *Foucault* suggests in a variety of works, *discourse* must be heard as

a synonym of his concept of *episteme*. Wikipedia – www.en.wikipedia.org/wiki/DISCOURSE
8. Rosalyn Diprose, 'A "Genetics" That Makes Sense', in R. Diprose, and R. Ferrell (eds), *Cartographies: Poststructuralism and the Mapping of Bodies and Spaces* (St Leonards, New South Wales: Allen & Unwin, 1991).
9. Hilde Lindemann Nelson, *Damaged Identities, Narrative Repair* (Ithaca, NY: Cornell University Press, 2001), p. xi.
10. Diprose, 'A "Genetics" That Makes Sense', p. 73.
11. Diprose, 'A "Genetics" That Makes Sense', p. 74.
12. R. Hubbard and E. Wald, *Exploding the Gene Myth* (Boston, MA: Beacon Press, 1993), pp. 37–38.
13. Lennard J. Davis, *Enforcing Normalcy: Disability, Deafness and the Body* (London: Verso, 1995).
14. N.A. Holtzman and T. Marteau, 'Will Genetics Revolutionize Medicine?', in *The New England Journal of Medicine*, 343.2 (2000), pp. 141–44.
15. Mary Johnson, 'Aborting Defective Foetuses – What Will It Do?', in *Link Disability Journal*, August/September 1990, p. 14.
16. Johnson, 'Defective Foetuses', p. 14.
17. Richard Lynn, *Eugenics: A Reassessment* (Westport, IN: Praeger, 2001), pp. vii–ix.
18. Lynn, *Eugenics*, p. ix.
19. We tend to construct the project of medicine as being inherently good and the epitome of the product of rational thought. It is both an *area of knowledge* (the *science* of *body systems*, diseases and treatment) as well as the *applied practice* of that knowledge. Yet in so doing we tend to forget about evil and bad things done by medical professionals and indeed the many errors found in medical systems and science. Biotechnology is constructed as a form of applied science, manipulating biological organisms to do practical things, providing useful products without necessarily recognizing the political dimensions and indeed vested professional interests found in its promotion.
20. For a discussion of this, see Christopher Newell, 'Christopher Reeve, Utilitarianism and Human Rights', in *Access*, 4.2 (2002), pp. 4–5. Also Goggin and Newell, *Disability in Australia*.
21. Michael Cook, 'Designer Babies? Don't Leave It to Bureaucrats to Decide', in *The Age*, 23 April, 2003.
22. David Wells, *Losing our Virtue: Why the Church Must Recover its Moral Vision* (Grand Rapids, MI: Eerdmans, 1998), p. 15.
23. After Goggin and Newell, *Disability in Australia*.
24. John Hockenberry, *Declarations of Independence: War-Zones and Wheelchairs* (London: Penguin, 1996), p. 257.

3

Life as Being in Relationship: Moving beyond a Deficiency-orientated View of Human Life

Martina Holder-Franz

In the beginning is the relation.[1]
　Martin Buber

I write this chapter from my experience as a pastor. As I have ministered to people with disabilities and their families, I have often been struck by the common tendency to define people with disabilities in terms of their deficiencies and perceived suffering. Rarely are the positive aspects of the life experiences of people with disabilities talked about. For the most part, the discussion focuses on charity and pity rather than friendship and relationships. When this happens, people with disabilities are easily sidelined from the community, and the important challenges that they bring are overlooked. In this chapter I will argue for an understanding of disability that is not determined by ideas of deficit or suffering, but rather focuses on the significance of relationships, both human and divine, as the essence of meaningful human living in the presence of God. I will argue that the human experience of disability is fully compatible with the good life if the good life is defined in terms of relationships. My claim is that when people with disabilities, those whom they choose as friends, their families, their communities and the institutions that comprise society begin to move from a focus on deficiency to one of relationships, the experience and perception of disability will change for the better. Human beings cannot be defined exhaustively by their biology or measurable qualities; there is always something more. It is that 'something more' that I want to explore.

Pastoring in the world of genetics

I recently had a conversation with a mother-to-be during which I started thinking about some of the issues that are highlighted in this book. The conversation went something like this:

'It's great to hear you're expecting your second child. How are things going?'

'I have to admit I'm a little unsure of things at the moment. You know I'm over 35 now and so I had some prenatal tests done. I didn't realize what that meant. Now we have to wait three weeks for the results. It's difficult – on the one hand we want to know if the child is healthy, on the other hand we wouldn't know what to think or do if it turned out there was something wrong. My mother has been very ill for years now, I know what illness and suffering mean. I have to admit I'm very scared. I don't know if we would have the strength to raise a handicapped child.'

Who can blame her for being afraid? Bringing up a child with a disability in a society within which disability is often not accepted and where social and relational support is not always forthcoming is not an easy task. In the course of my pastoral work I often come into contact with women and families who share with me their anxieties and wishes before, during or after their pregnancy. Part of that experience relates to prenatal testing. Such tests are now standard in the Western world, and many, perhaps most, women and couples have no objections to having them. Indeed, people often see such tests in a positive light, in that they are perceived as an affirmation that their situation (i.e. their age and environment) is being taken seriously and that they are being properly looked after and well-advised medically. Those who want to have and look after children hope that their physician will be able to give them a 'good report': 'Yes, Dr X said everything is all right. *Now* we're thrilled about having our baby.' The idea that one can only be courageous enough to be thrilled about the baby once one is assured that it is not disabled is interesting, and is discussed in more detail elsewhere in this book.

However, often in my experience nothing much is said about the two to three weeks between taking the tests and receiving the results, or of those parents who receive the news that their child is disabled or who get an uncertain report from their doctor. Usually women in this situation are expected to then take a number of further tests. For some this will result in the discovery that the child is not disabled (such tests have a fairly high margin of error). I have ministered to a number of families who, after much thought and consideration, have decided to continue with their pregnancies in spite of the tests indicating that their babies would be disabled, only to discover that their child was born without any form of disability.

Life-and-death decisions

There are, however, other parents who receive a diagnosis that their unborn child has a disability, who are then faced with very difficult decisions. People in this situation find themselves tormented with decisions over life and death to which earlier generations were not subjected. Like the woman in the pastoral encounter highlighted above, people find themselves placed under enormous social, economic and relational pressure. To have the child would be very difficult, but not to have the child brings with it other problems. It is not my intention here to discuss the moral and ethical objections to abortion. Whatever one's opinion on this, the literature indicates that the experience of abortion frequently leaves behind deep psychological, relational and sometimes physical traces on people's lives. As a pastor, it is clear to me that people seldom get over the experience of abortion without some form of enduring difficulty.[2]

A discontinued pregnancy is experienced as a deep crisis in the life of a person or in a partnership, one which causes problems and takes a long time to overcome. It is much more than a medical procedure. It is inevitably a fragmentation of relationship. The mother is connected physically to the child and so already has a relationship of sorts with the developing life inside her. If on the physical level the pregnancy is discontinued, one might suppose that the breaking of the relationship with the developing child will leave behind some sort of psycho-relational trace or scar. It is true that we have a much more liberal and accepting view of abortion within society. However, these changes in perspective and opinion need to be brought into dialogue with the actual questions which many people have regarding what it means to live a good and meaningful life and how we deal with issues of guilt and forgiveness within our pastoral encounters.

Caring for disabled children

The point I want to emphasize is that encountering disability in a pastoral context is complex and often messy. In my experience, when parents are faced with the types of serious decisions regarding life and death I have highlighted thus far, they simply lack the resources to know what to do. Rarely is it the case that parents would reject a child with disabilities on a basic level. Their problem is not necessarily with disability *per se*, but rather that they feel they simply don't have the emotional, social or economic resources to cope with the possible complications of having a disabled child. They are afraid. The idea of having a child with a severe handicap fills them with dread, and they worry that this experience would stretch them to such a degree that they wouldn't know how to cope. They fear that they wouldn't know how to go about accepting a child with a handicap. 'How could I live with such a child?' A pregnant mother in my church, who already has three small children, was all of a sudden confronted with the possibility that her fourth child might have Down's syndrome. They rang me up in near despair, fearing that they wouldn't have the strength to welcome the child, even though they knew it would probably be the right thing to do. These situations are difficult and require much more than simplistic responses.

The quest for perfection and the place of the disabled in society

The danger of eugenics

However complex such pastoral situations may be, they do raise important questions about the type of society we have chosen to create. Why is it that we have such a problem with disability? Why is it that we seem constantly, both as individuals and as a society, to prize the idea of 'perfection' and balk at the idea of difference? Questions such as these raise for me the difficult question of eugenics.

Through the discussion on prenatal testing and abortion I have tried to draw out some obvious pastoral dynamics. However, there is a wider debate beyond what we have examined. Ursula Stinkes warns that the increasing liberalization of the regulations regarding abortion will inevitably devalue the life of people with disabilities.[3] The more commonplace it becomes for people to choose whether they want to have a particular baby or not, the less chance there is that those perceived as the weakest will be chosen. The pressure will grow on parents-to-be to avoid 'weak' babies.

An interesting contribution to this discussion was given by an exhibition in the German Museum of Hygiene in 2001. The institution had played an infamous role during the 1930s in its campaign to propagate the ideology of Aryanism. The title of the exhibition was, 'The imperfect human – concerning the right to not be perfect'. Through pictures, documents and different events the public were invited to consider their definitions of what it means to be 'normal' or 'healthy'.[4] Dieter Mettner in his important contribution to the sociohistorical development of the concept of normality writes in regard to the nineteenth century:

> during the phase of the beginning of industrialization in the 19th century, and in the social tension caused by dwindling resources in the final years of that century, there emerged under the impression of social-Darwinist theory ominous perspectives, during which the health and population policies of the time got caught up in a whirlpool of biological control programs . . . The idea of eugenic measures to avoid the 'social ballast of those with hereditary defects', and the breeding of genetically healthy persons, developed in the 19th century. In social-Darwinism the laws of evolution as Charles Darwin saw them in his observations of nature were transferred to human development.[5]

In Mettner's opinion, after 1945 there was not sufficient debate within the scientific community about this historical inheritance.

> After 1945 the internationally common and not-yet-loaded terminus 'human genetics' was used instead of the earlier label 'hereditary research'. A similar social-Darwinist eugenic ideology appears more and more today, however, in the robe of genetics, an ideology which laments the

increasing genetic and biological degeneration of humanity . . . and out of a concern for the quality of the gene pool demands that people be made biologically fitter for the wear and tear of the 21st century . . . The goal of this postmodern, technicistic, humanist project is a genetically constructed idea of 'normality' which regards everything else other than this normality (the genetically inferior) as an avoidable risk. That which doesn't meet the desired criteria of quality, the un-normal, can now, in contrast to the euthanasia excesses of earlier times, be disposed of in a totally unspectacular and clinical fashion by means of molecular-biological selection or prenatal intervention, or be avoided by the already practiced, new reproduction technologies.[6]

Common contemporary arguments citing prenatal testing as having to do with the avoidance of suffering and certain illnesses, when in reality they involve the destruction of disabled foetuses, remind one uneasily of these patterns of thought, particularly when incorporated in discussions of the analyses of cost-effectiveness.

If this is so, then reflection on the pastoral experiences that opened this chapter have brought us to an interesting revelation about aspects of our society that we might not otherwise notice and with which we may well be uncomfortable. We need to keep a critical eye on developments in society; it is important that such eugenic tendencies be seen and openly named for what they are. Not to do so puts people with disabilities in a very tenuous position. It is my opinion that we must be aware of and fight against the tendency to create a 'myth of normality' that reduces people to nothing more than their biological characteristics and qualities. There is something more than disability which makes up the essence of the person. Indeed, being disabled is only one of the different forms and experiences in which a person exists. What then is the essence of a person? What is quality of life? There are many people with disabilities who strongly advocate the position that being disabled and having good quality of life are not mutually exclusive. Indeed, they emphasize that disabled people contribute something important to others and to society in general and they compel a society to review its values and goals and consider anew the vulnerability of life.[7]

Quality of life as 'being in relationship'

This suggestion brings me to my central thesis: *that which truly constitutes quality of life is being-in-relationship*. Relationships take place on very different levels, but it is in and through them that we discover what it means to be a person. A relational perspective on the human person means that they are not defined by ability, but rather by their availability for relationships. Personhood is thus sustained even in the presence of severe limitations. This model challenges the types of deficiency-orientated definition of human life that I highlighted at the beginning of this chapter.[8] Furthermore, an approach that takes into account different levels of relationship can sharpen the perception of the resources each human being has for developing relationships regardless of his or her disabilities.

This is an encouraging view. Moving from a deficiency-orientated perspective to a resource-orientated view opens up challenging new possibilities which not only make life more tolerable but also deeply rewarding. A resource-orientated view of human life, as I would define it, contends that every person has something valuable within themselves which can be shared. If, for example, I visit a home for people with learning disabilities where someone is introduced to me, and I am immediately informed of what they cannot do and what will not be possible despite all our efforts, I will have a totally different idea of the person than if I am told about their favourite pastimes, about the friends they have, and about the things they can do and enjoy doing despite perhaps extremely severe limitations.[9] If the person is introduced as a problem, the chances are I and others will act accordingly. But if they are introduced as a person with whom I might desire to develop a new relationship, everything changes. In this approach it is important that the view of mental disability as a deficient part of the essence of a person is rejected, and that the person is not defined as a deficient individual but as a person in relationship who has a disability.[10] This approach, which regards being in relationship as the essence and real quality of life, doesn't mean that making decisions is any less difficult or complex, but it does significantly alter the starting-point for all discussion: it is relationships and not particular abilities or dis-abilities that are meaningful and definitive for our understanding of the good life.[11]

The grounding of such relationships is of course in God. Such a relational ethic is found in the work of Emmanuel Levinas and Martin Buber. For example, Buber argues that

> The person who says 'You' doesn't possess anything, has nothing, but that person stands in relation to someone . . . a relationship is mutuality . . . I cannot experience the person to whom I say 'You'. But I have a relationship to that person . . . All essential life is relationship . . . Love is responsibility of an 'I' for a 'You'. Creation reveals its possibilities in relationships . . . A person who meets a 'You' becomes an 'I'.[12]

It is as we encounter one another as 'Thou' rather than 'It' that we discover who the other is and what it means to be in relationship with them. Here we discover what Emmanuel Levinas has described as 'the epiphany of the other person': 'her countenance carries its own meaning within itself, one which is independent of any meaning the world outside gives it. The other person doesn't just approach us through our context: she is meaningful in herself.'[13]

In relationships we find one another and discover ourselves. Much of the discourse around disability is carried out in the mode of 'it'. The epiphany of the other person is often not available within the moral discourse as it is normally developed. The relational model I am emphasizing takes seriously Buber's identification of the relational significance of 'thou' and Levinas's 'epiphany of the other person', and seeks to embody this way of being in forms of caring for people with disabilities and their families which allow for the development of meaningful relationships and the making of decisions to be framed in relational rather than conceptual terms.

A Christian perspective

The question is yet to be asked, what is quality of life from a Christian point of view? Quality of life in the Judaeo-Christian tradition, as I see it, can be summed up in one word: love. Obviously love has everything to do with being in relation. Irrespective of our particular theological approaches, they all lead us to the same conclusion. Creation theology, the *Imago Dei*, covenant theology, the central importance of the incarnation, and the Trinity all find their central focus in loving relationships. These approaches differ from each other, but in essence they all have to do with relationships. The text that I find most important for the argument I am developing in this paper is the question that Jesus was asked by the scribe in Luke 10.25–28: 'What must I do to inherit eternal life?' Jesus asks the scribe, who knows his Bible well, what the scriptures say. The scribe cites from Leviticus: 'You shall love the Lord your God with all your heart, and with all your soul, and with all your strength, and with all your mind; and your neighbour as yourself'. And Jesus said to him, 'You have given the right answer; do this and you will live'. This text is all about relationships: the relation to God, to one's neighbour, to oneself. Quality of life in the Judaeo-Christian perspective is reached where these three relationships are adequately taken into account.[14] Jean Vanier takes this approach further in his thoughts on John 13.34–35:

> In the Law of Moses, the Hebrews were called to love God with all their soul, heart, mind and strength and to love their neighbours as themselves. Here Jesus is calling his disciples not only to love others as they love themselves but to love as he – Jesus – loves them. That is what is new. He is creating a holy sacred covenant between them. They are called to live in communion with each other, to share with one another, to serve one another in simple acts of love and caring, never judging or condemning but forgiving.[15]

The most important words in the Judaeo-Christian faith, the greatest commandment, are words about relationships. That doesn't mean that the physical aspect has no role to play. One's relationship to one's neighbour and to oneself is only possible within the form of a body. It does, however, mean that the biological, the physical, should not be overemphasized. Our essence, the quality of our lives, is not bound to our perfect physical health.

A caring community

In closing this chapter I would like to return to my conversation with the pregnant woman from my parish. Part of my job as a pastor and a Christian counsellor is to listen carefully to what this woman has to say. However, another part of my task relates to encouraging her in her relationships with her child, her family and her community in order to strengthen her resolve that the acceptance of a child, even if it is disabled, makes sense and is good. With that there is also an obligation on my part and within the Church in general to accompany her and all who find themselves in a similar situation, to support her and to look after her in order that she can find the strength and the will to build a relationship with her disabled child.[16]

If parents-to-be can speak with other parents who look after a disabled child, if they experience in their family, their circle of friends or in their church that people are willing to stand by them and support them, then perhaps they may start to see the acceptance of a disabled child become a possibility. Churches have a special responsibility in this area because they cannot postulate demands and wishes on others without being fellowships which are willing to live out their principles and be committed to the relationships in which they believe. Worship together with mentally or physically disabled people shouldn't be simply a spoken ideal in the church but an everyday practice, however challenging that may be.[17]

Conclusion

The quality of life-in-relationship is of course not something that comes to us naturally, but rather is something that grows out of faith in God. The belief in a God who knows who I am, who addresses me personally, means that I have a relationship which no one and nothing can take from me, not even the experience of disability. The Christian tradition must emphasize again and again that freedom from bodily disease or disability is not a prerequisite for relationships which make up the essence of the good life. Faith in Jesus Christ opens up our eyes to see that God's 'Yes' to human beings is not just for the physically strong and fit but also for those who are weak and disabled, the sick and fearful.[18] This 'Yes' from God is not just a dutiful toleration of the 'other' but is a powerful word spoken out of love. It is this love that Jesus commands us to practise, and it is this love that offers fresh perspectives and new possibilities for living with and for people with disabilities in a world that, as we have seen, is far from simple.

Notes

1. M. Buber, *I and Thou* (trans. Walter Kaufmann; Edinburgh: T&T Clark, 1970), p. 69.
2. See the results of family therapies and the work of Virginia Satir. In the family reconstruction process every pregnancy, including those which end in the loss of a child after a miscarriage or an abortion, must be recognized. Virginia Satir, *Kommunikation, Selbstwert, Kongruenz* (Paderborn: Junfermann, 1990). See also A. von der Schlippe and J. Schweitzer, *Lehrbuch der systemischen Therapie und Beratung* (Göttingen; Vandenhoeck and Ruprecht, 2003).
3. Ursula Stinkes, *Menschen mit einer geistiger Behinderung* (Göttingen; Vandenhoeck and Ruprecht, 2003), p. 57.
4. Dieter Mettner, *Der imperfekte Mensch: Vom Recht auf Unvollkommenheit* (Berlin: Hantje Cantz Verlag, 2001).
5. Mettner, *Der imperfekte Mensch*, p. 26f.
6. Mettner, *Der imperfekte Mensch*, p. 28f.
7. Interview with Simea Schwab, a severely physically handicapped theologian in Switzerland, who regards her calling as a theologian in this light (15th April 2005, Kerzers, Switzerland).
8. J. Weiss Block, *Copious Hosting* (New York: Continuum, 2002), pp. 47ff. She speaks about 'oppression through domination by non-disabled people'.
9. See, for instance, Stanley Hauerwas, *Resident Aliens* (Nashville, TN: Abingdon Press, 1989), p. 93.
10. Stinkes, *Menschen*, pp. 40f.
11. See, for instance, the virtue ethics debate, i.e. Alisdair MacIntyre and Iris Murdoch, in Alisdair MacIntyre, *After Virtue: A Study in Moral Theory* (Notre Dame, IN: University of Notre Dame Press, 1981).
12. M. Buber, *I and Thou*, pp. 79ff. (trans. here by Dan Holder).
13. See Huizing, *Aesthetische Theologie, der erlesene Mensch* (Stuttgart: Kreuz Verlag, 1983), pp. 200f.
14. I am aware that there are various attempts to define quality of life. Oliver Tolmein criticizes different approaches in his book, *Wann ist ein Mensch ein Mensch: Ethik auf Abwegen* (Munich: Hanser Verlag, 1993), pp. 142ff.
15. Kathryn Spink, *The Miracle, the Message, the Story: Jean Vanier and L'Arche* (London: Paulist Press, 2006), pp. 189f.
16. It is not a coincidence that in the gospel of Luke the question of what is most important is immediately and practically illustrated by the parable of the Good Samaritan.
17. See John Swinton, *Building a Church for Strangers* (Edinburgh: Contact Pastoral Trust, 1999), pp. 11f.
18. Settimio Monteverde points out that Isenheim Altar (now in Colmar), with its vivid images of the suffering Christ, was originally intended to be seen by those who were sick and dying, and for those caring for them, as

a consolation that they were not rejected by God, but that Christ himself suffered with them. M. Mettner and R. Schmitt-Mannhart (eds), *Wie ich sterben will* (Zürich: Auflage Verlag, 2003), pp. 293f.

4

Arguing about Genetics and Disability
Tom Shakespeare

This article presents a dialogue between two hypothetical characters, in order to rehearse some of the main arguments against prenatal screening, and highlight some of the problems with these arguments. The author's own views are a composite of both characters' positions.

Opponent of prenatal screening: Genetics is just the same as eugenics. It's about eliminating 'lives unworthy of life', which is what the Nazis did.
Advocate of prenatal screening: I assume you're not implying that geneticists are Nazis, because that's untrue and offensive. But when it comes to eugenics, it depends what you mean by the word: it is notoriously difficult to agree on a definition. The main differences between early twentieth-century eugenics and the present practice of prenatal screening is that pre-1945 it was a matter of state policy, and it often involved coercion. In present-day Western countries, pregnancy termination is the free choice of individual women and men, within the parameters of the law.
Opponent: Well, I'm not opposing a woman's right to choose. If a woman decides not to continue with her pregnancy, then that's up to her. But choosing whether or not to be pregnant is a different choice from deciding which foetus to be pregnant with. I support that first right to choose, but not the second. I think it is discriminatory to make that choice on the basis of the characteristics of the foetus.
Advocate: I'm not sure you can separate the issues of 'being pregnant' and 'being pregnant with a particular foetus'. Consider the hypothetical case of an unmarried 16-year-old girl who becomes pregnant. Her decision as to whether to continue and become a single parent might be different in the case of a potentially non-disabled baby than it would be in the case of a potentially affected baby. She might decide she could cope with the former situation – knowing that day care and all sorts of other support might be available – but that she could not cope with being the single parent of a baby who has high support and care needs. And surely it is counterintuitive to allow women to exercise their right to choose termination for social reasons – such as failed contraception, or a change of heart, or being too old or young, or having too many children – but to deny it in the cases of diagnosed serious foetal impairment?
Opponent: OK, I allow your point that choice is an important principle. But

I don't believe that women and men are exercising free and non-constrained choices in practice, for several reasons. First, they are not given full information. Sometimes they are not given proper clinical information about the particular impairment that has been diagnosed. Usually they are not given full social information on what it's like to have that impairment, and the quality of life implications. They are rarely told about the psychological impacts of termination of pregnancy. And they are never provided with the perspectives of disabled people themselves, who are the real experts on being disabled. Second, doctors and other professionals are biased against disabled people. They are ignorant about disability. They think disability is a tragedy to be avoided at all costs. They do not counsel non-directively. They believe screening is a good thing, and this influences their patients. Third, the clinical context influences the choices made: for example, making a test available implies the desirability of that test. Antenatal testing is like a conveyor belt, and many women are not given the time and information to make an informed decision. Finally, society is increasingly blaming women for not having tests or not having terminations. For all these reasons, there is no real choice at the moment, and women are not supported to continue with pregnancy if they want to do so.

Advocate: I accept your argument. Choice at the moment is rather limited. Society needs to ensure that resources are invested to enable women and men to make fully informed choices. If we managed to achieve that, then you would have no reason to prevent women having the choice of terminating a pregnancy affected by impairment, if they wanted to, would you?

Opponent: I'm not sure. I don't like the idea of people trying to avoid the birth of disabled people. It's like saying disabled people aren't worthy of life . . .

Advocate: That's not necessarily the case. I can think of four reasons a woman might choose to avoid the birth of a disabled child. One is because they don't think disabled people should exist, or because they think that disabled people aren't worthy of life. Another is the argument that society should not have to pay the costs of supporting disabled people. Both of these seem to me to be morally dangerous. I would join you in opposing them, and I would call them 'eugenic' reasons. But the second two reasons seem to me to be important. One is that impairment involves suffering and physical difficulties, and it is unfair to bring people into the world to suffer. The second is that it is often difficult for the parents and siblings of disabled children. There is sometimes a very negative impact on the whole family. Relationships break up, and brothers and sisters may become neglected or resentful.

Opponent: I am glad you agree that your first two reasons are oppressive and should be opposed. But I think you've missed the point about disability. Both your second two reasons are not really about what it's like to have an impairment. They are about the way society treats someone with impairment. Disabled people say their real problems are discrimination and prejudice in society. That's what makes life difficult for disabled people and their families. We should remove the social factors that cause suffering and isolation, rather than remove disabled people from the world. We need social engineering, not genetic engineering!

Advocate: OK, fair enough, I agree we should try and change the world, although I reckon that it might be much harder than you think to change some of these social problems. But surely not all impairments are the same. There are some impairments which are invariably very difficult. Babies die in their first year of life, or people die before their twentieth birthday, or people live very difficult lives with limited consciousness and self-awareness, or else with extreme pain and physical difficulties. However much you change society, surely these problems will remain and should be avoided if possible?

Opponent: Well, that's just it: I don't think it's up to us to try and remove impairments from the world. Impairment is a fact of life – after all, we're all going to die. Being alive involves suffering. We shouldn't be playing at God.

Advocate: Well, impairment may be inevitable, but that doesn't mean we don't have a duty to try to minimize it, especially when it is very severe and debilitating. After all, we agree on some tactics for removing impairment, such as the vaccination of children, or mine-clearance, or looking both ways when we cross the road. Nobody would have a problem with impairment prevention, would they?

Opponent: I think there's a difference between impairment prevention and removing people with impairment from the world. And where do you draw the line? If you are giving women the right to choose, does that mean the right to terminate pregnancy on the basis of the sex of the potential child, or perhaps sexuality or intelligence? If you are going to be consistent about choice, then why stop at impairment?

Advocate: There is no requirement to be either totally pro-choice, or totally anti-choice. Ronald Dworkin argues against the notion of 'foetal interests' and believes that termination of pregnancy is not immoral. However, he argues that this does not mean that termination of pregnancy is a morally insignificant act. It involves halting life once it has started, and should not be entered into lightly. Because termination of pregnancy is morally significant and important, it should be chosen only when the alternative would be much worse for the parents or potential child. For this reason, termination of pregnancy on grounds of personality characteristics – for example gender and sexuality – should be avoided.

Opponent: It's all very well resorting to philosophical arguments. But the fact is that I might not have been born if these selective termination techniques had been available to my parents' generation . . .

Advocate: Your statement has immense emotional weight, but it does not make sense. Saying 'I would not have been born' is not logical. The point is that you were born. Prior to your birth, there was no 'I'. Only after your birth was there an 'I'. Souls do not wait in limbo before birth, being prevented from coming into the world by particular acts of contraception or termination of pregnancy.

Opponent: OK, I accept that, if you want to be pedantic. What I meant was, these techniques stop disabled people in general from being born.

Advocate: But that's not strictly true either. You have accepted that termination of pregnancy is morally acceptable, presumably on the basis that up to a certain stage of pregnancy – say 24 weeks – there is no 'person' involved, just a 'potential person'. Termination of pregnancy stops a collection of cells developing further. It does not stop a disabled person being born. However, the effect of selective termination may be to reduce the number of disabled people in the world.

Opponent: That's exactly what I mean. Selective termination reduces diversity. And what's more, terminating foetuses affected by the same condition as me is a form of discrimination against disabled people. It's a judgement on me and on my life. It will lead to more prejudice against disabled people.

Advocate: I am not sure that there is any evidence that selective termination of pregnancy increases prejudice against disabled people. In China, for example, there is a strong eugenic policy, but there is also increasingly good provision for disabled people. And the fact that we take a sugar-lump inoculation against polio does not cause discrimination against people with polio. Prevention and support are not incompatible.

Again, I can see the emotional relevance of your feeling discriminated against because of a screening programme designed to eliminate your impairment, but I don't think it is just or rational. After all, let's say you got your way, and out of respect to you, society decided to prohibit selective termination on the basis of your impairment. What would you say, in twenty years time, to the person who was born with the impairment, when they complain that you stopped the technology being used to prevent the birth of people with that impairment? Why should they suffer because the idea of impairment prevention makes you unhappy or feel discriminated against?

Opponent: OK, maybe I shouldn't have talked about discrimination. But I notice that you haven't dealt with my argument about diversity. I still think that selective screening could go too far. I accept that people should have reproductive choice, but I don't want to see a world in which all impairment has been eliminated. We should value every individual, and we should support difference. I want to see these technologies carefully regulated; I want to see informed and supported choices; and I think we should recognize the contribution which disabled people make to the world, and their right to be a part of it. If we challenge the prejudice and fear which surrounds disability, prospective parents would be less likely to feel that termination was the only answer.

Advocate: I don't have any problem agreeing with that. I can see why you feel insulted and denigrated by the rhetoric which supporters of genetic intervention sometimes adopt. I think we should be careful not to say disability is invariably a tragedy, and I think we should try to reduce the discrimination which often makes the lives of disabled people so much more difficult. But I do believe that we should offer women and men access to screening information, and give them a free choice about whether to continue with pregnancy.

Perhaps there's also a difference between testing which involves families who already have a history of genetic conditions, and who know what they involve, and screening which extends genetic intervention to the whole population. Screening is often introduced on the basis of cost-benefit calculations about avoiding the birth of disabled people. It seems to be where genetics comes uncomfortably close to eugenics. Perhaps we should be in less of a hurry to introduce the latest tests, or extend them as widely as possible. Biotech corporations might be keen for us to take advantage of these technologies, but if we cannot guarantee that people will be informed and supported to make the best choices, then perhaps it is too risky to push ahead with this type of screening.

Notes

This chapter first appeared as an article in *Interaction*, 13.1 (2000), pp. 11–14.

Thanks to Tom Shakespeare for permission to use the article in this book.

Part 2
Eugenics and the New Genetics

5

To Form a More Perfect Union: Mainline Protestantism and the Popularization of Eugenics

Amy Laura Hall

I began to suspect serious links between mainline Protestantism and American eugenics around the time I learned of Christine Rosen's extensive research on the subject. She readily shared early drafts of her *Preaching Eugenics*, has gone digging through her copious notes to answer my questions, and has been a true gift to me during this project. I must express here my debt and appreciation. Sharon Leon has generously shared her own detailed research on the contrasting Roman Catholic opposition to eugenics and has answered many questions along the way. I owe Dennis Durst for both contacts, as he so graciously introduced me to Sharon's and Christine's work. Dennis has also shared his fascinating research on the role of American evangelicals in eugenics.

Figure 5.1 How many? How healthy?[1]

The first urgency is to know the axioms of eugenics. We are not even well educated nor modern if we have no bowing acquaintance with its larger truths.

(Revd Phillips E. Osgood, St Mark's Church, Minneapolis, 1926)

Phillips Endecott Osgood's sermon 'The Refiner's Fire' won top honours in the first of a series of well-received sermon contests sponsored by the American Eugenics Society (AES). Using Malachi 3.3 as his text, Osgood admonished the 'temporary guardians of a miraculous gift' to respect their 'partnership with God to keep it pure'! 'The dross must be purged out' – and eugenics was the means for purging. The editors of the ecumenical journal the *Homiletic Review*, reprinting the sermon in 1929, registered their unequivocal assent: 'A contest is worthwhile which evokes so excellent a sermon. Eugenics is an approved thing. It is no longer on trial.'[2] Twenty years later, an avid Protestant eugenicist named Paul Popenoe would put the matter more subtly: the young Presbyterian, Methodist and Episcopal men and women reading the YMCA/YWCA-sponsored magazine needed to take on the civic responsibility of 'Surveying the Chances'. Citing his own study of children in the Sonoma State Home, he warned that 'feeble-minded' families were reproducing at a higher rate than those able to provide 'the best start'. Reminding his mobile, collegiate readers that 'cities always live as parasites on the rural areas', Popenoe emphasized their duty to consider patterns in 'the nation's birth-rate' when choosing their mates and planning their families.[3] The picture above linked Popenoe's article with one by Helen Southard entitled 'Planning Parenthood on Campus'. By the mid-twentieth century, the matter was largely settled among mainline Protestants: 'The Christian asks: how many; how healthy?'

The history of the eugenics movement in the United States is seldom brought to the public eye. When popular historians do present the story, eugenicists are usually safely on the other side of a wide intellectual and cultural gap. Their science was faulty. Their ideas were blatantly racist. To quote one Public Broadcasting System (PBS) narration, the 'horrors of institutionalized eugenics revealed in Nazi Germany . . . doused [American eugenics] entirely as a movement' after the Second World War.[4] Arguably one of the trickiest tasks of narration is the one facing the writing staff at Cold Spring Harbor Laboratory. The lab recently celebrated '100 Years of Genetics' in 2004, reflecting a history that extends from Charles Davenport, lab director and mastermind of the Eugenics Record Office (ERO), to current chancellor James Watson, co-discoverer of the double-helix structure and advocate for 'making better human beings' through genetic engineering.[5] In May 2002 David Micklos, director of the lab's Dolan DNA Learning Center, marked the 75th anniversary of the landmark eugenic sterilization case, *Buck v. Bell,* with a special online article, 'None without Hope: Buck *vs.* Bell at 75'.[6] That original test of Virginia's sterilization laws went to the US Supreme Court in 1927, where Oliver Wendell Holmes, writing for the court, rendered the following:

> It is better for all the world, if instead of waiting to execute degenerate offspring for crime, or to let them starve for their imbecility, society can

prevent those who are manifestly unfit from continuing their kind. The principle that sustains compulsory vaccination is broad enough to cover cutting the Fallopian tubes. Three generations of imbeciles are enough.[7]

With characteristic clarity, Justice Holmes thus linked several key eugenic concepts. In order to prevent individual suffering, the state may compel the prevention of certain 'kinds' of individuals. As an effective inoculation against degeneration, crime and imbecility, the social body may 'vaccinate' itself against the deleterious or parasitic 'unfit'.

Using subtle rhetorical cues, Micklos embeds this egregious story within the longer history of genetics at Cold Spring, distinguishing between the lab's dubious past and its promising present. Although Cold Spring's home team, led by Davenport, paved the way for the sterilization of Carrie Buck after the birth of her daughter, Vivian, their science was sullied by the 'biblical concept that "like breeds like"'. Impure, their scientific methods were eventually discredited by a more accurate strand of genetics; coercive, their political methods were untenable after the Nazi atrocities. As evidence of the state's mistake, the site features a link to Vivian Buck's first-grade report card, telling us that this supposed third-generation imbecile eventually made the elementary school honour roll. Micklos here brings the reader back to the point of his title. As it turns out, no lineage is 'without hope', because the new science of genetics is revealing a complicated combination of factors – factors that might 'predispose a person to autism' or 'predispose to genius' so that 'one can never predict where genius will arise'. Micklos situates the eugenics of the past on the other side of a chasm, distant from now-chastened politics and a science whose backward, biblical myopia has been duly corrected. The piece concludes that the Buck girls would likely wish us to take this lesson of 'hope' with us into our 'Brave New World'.[8]

Figure 5.2 A burden on the rest [9]

This chapter is in part my attempt to tell a different story. One sign from the American Eugenics Society's Fitter Family fairs from the first half of the last century names a core assessment that still holds purchase: 'Some people are born to be a burden on the rest'. This gauge applied then to many kinds of difference; those who would judge scanned the horizon for those who variously did not 'fit'. Depending on the region, the signs of unfitness included poverty, disability, race and religion. I hope to complicate the standard narration of eugenics past and present by suggesting that this core assessment led to an arsenal of biotechnological tools to plan, evaluate and enhance children and to measure the worth of a given family tools that today have become standard parental and political equipment. I will also suggest links between current hopes for genius and past attempts to vaccinate the social body against the menace of poverty, disability and deviance. As individual parents navigate the strand of genetics that supplanted the science of Davenport and his ilk, they are choosing in rising numbers to terminate pregnancies that show signs of genetic difference choosing, in the majority of cases, to terminate for conditions ranging from physical disease to mental disability to gender ambiguity. At the same time, citizens in the United States view with increasing scepticism public spending on the supposedly indiscriminately bred children of poor African-American mothers as well as on the 'huge' families of recent Latino immigrants.[10] The cultural context in which individuals make what are increasingly seen as purely 'personal' decisions and in which a society makes what are often deemed purely pragmatic decisions is shaped by the powerful rhetoric of eugenics.

The pregnant body, the social body, and the burden of certain 'types' of babies are all culturally loaded in ways that reflect the vast movement in the past century in the United States to popularize eugenics. The quest to craft a more perfect union through 'fewer and better babies' is alive and well.[11] *Pace* Micklos, *pace* PBS, *pace* the tale often told, many eugenic ideas have jumped the gap from yesterday to today, bridging the chasm between overtly coercive eugenics and purportedly voluntary parental and social responsibility in the land of the free.[12]

The hypothesis of my narration is that eugenics gained popular support in large part through the endorsement of those mainstream and progressive Protestant spokesmen and women. Well after Henry Ward Beecher, self-declared 'cordial Christian evolutionist', endorsed from his pen and pulpit the social use of Charles Darwin and the social Darwinism of Herbert Spencer, mainline Protestantism continued to accept the thoroughgoing relevance of eugenic ideas.[13] From Paul Popenoe's *Intercollegian* and *Ladies' Home Journal* articles encouraging white middle-class men and women to replenish the race to the current United Methodist endorsement of 'responsible' parenthood in the UMC Social Principles, mainstream Protestantism has lent legitimacy to a trajectory of discriminating reproduction. Leading eugenicists in the United States used their own white, middle-class, literate, wholesome, productive, patriotic, native-born Protestant families as the standard by which other families would be measured and judged. The eugenics movement was germinated in a relatively elite, academic version of scientific racism from the previous century, but it took root in the heartland of America, arguably as a result of two primary forces: clergy eager to remain relevant in an era when other professionals, and scientists in particular, were gaining ascendancy; and middle-class laity eager to establish themselves as good, wholesome parents and productive, responsible citizens.

The 'born to be a burden' sign from the AES Fitter Family fairs warned with two intermittently flashing lights that of all persons born in the United States – that is, one 'every 16 seconds' – a mere 4 per cent, or one 'every 7½ minutes,' is 'high-grade', able to 'do creative work and be fit for leadership'. A third flashing light underscored the associated economic toll: 'Every 15 seconds $100 of your money goes for the care of persons with bad heredity . . .' That this display carried such rhetorical weight – that across the north-east and the heartland, farmers and shopkeepers and homemakers and minister's wives had their own and their children's heads and limbs measured and their extended family-trees mapped for taint or purity in order to be identified among the 4 per cent of 'high-grade' people – is a sign that the AES knew well their constituency. That this display evoked among mainline Protestants hopeful aspirations rather than holy offence is an important part of the story of American eugenics.

Holy husbandmen – 'Interpreting the historic faith in a modern world'

Figure 5.3 Eugenics tree [14]

One symbol of the eugenics movement in the 1920s and 30s was a 'Eugenics' tree, with roots branching out to tap the 'many sources' from which the eugenics movement drew in order to become a 'harmonious entity'. Through the 'self direction of human evolution', eugenicists hoped to cultivate a tree that would flourish, bearing only good fruit for the future. It is worth noting that 'religion' was relegated to a root well off to the far right, one of the furthest roots from the main trunk of the tree. Biology, psychology, statistics, mental testing, history, geology, law and politics all had roles to play, with religion seemingly squeezed in almost as an afterthought – after sociology, to be precise. Yet the imagery of human advancement, hope and flourishing fruit was rhetorically potent, owing in part to echoes of the same biblical faith preached by Davenport's Congregationalist father, and the movement needed clergy to keep the echoes resonant.[15] Through the formation of the AES, eugenics leaders signalled that it would not be sufficient merely to keep careful measurements and records of the unfit and the fit, the impure and the pure, through the ERO. They needed to capture the imaginations of the citizenry. Religion was perhaps the root closest to the ground, and the AES found there clergy eager to prove that they were on the modernist side of the modernist/fundamentalist rift. Through cooperation with the AES, the YMCA and the ASHA, through thoughtful reviews and provocative sermons, clergy took up their calling in that 'great field of usefulness' that supported Protestant eugenics.[16] The sociologist, the psychologist, the anthropologist and the social worker were on the ascendancy, and mainline pastors throughout America were determined not to seem obsolete or, even worse, backward. As one reads the 'modern' attempts to prove a bowing acquaintance with eugenics, it is not difficult to surmise why mainline laity were unable to resist the allure of proving their own families fit at the expense of others. Faced with the challenge of remaining relevant and seeming well-educated in a modern world, many mainline Protestant clergy serving parishes and academe, in cities and in the country, did nothing less than capitulate.

In her meticulous treatment of the key religious players in the popularization of eugenics, Christine Rosen detects this as a 'clear pattern': 'The liberals and modernists in their respective faiths those who challenged their churches to conform to modern circumstances became the eugenics movement's most enthusiastic supporters'.[17] Rosen's conclusion is irrefutable. To cite one contrast, the Roman Catholic Church – with marked consistency from the grassroots to the Vatican resisted laws against sterilization as well as the mindset behind the movement, while some of the most hearty supporters of the AES found happy soil in the Anglican and Episcopal churches. The transatlantic links between the latter church allowed eugenic theologians such as Canon Charles Kingsley to influence priests such as Karl Reiland (in New York City) and Walter Taylor Sumner (in Chicago), who were eager to prove themselves legitimate fruit of the *forward*-thinking branch of the Apostolic Church. One article in the *Christian Century* from 1924 is telling. Reporting on a conference on the Church and science held by the aptly named 'British Churchmen's Union for the Advancement of Liberal Religious Thought', the piece relates at length the words of Oxford University professor and lay Anglican J.S. Haldane. In the quotation, Professor Haldane insists that people of faith 'cannot afford to be hampered by unintelligible beliefs which are mainly materialistic accretions of Christianity and which greatly weaken its influence on those who are worth influencing'. Distinguishing 'religion itself' from these accretions, Haldane warns that 'any shirking of the questions involved or cowardly sheltering behind mere traditional authority is fatal'.[18] The desire to remain in the good graces of 'those who are worth influencing', namely, (apparently) those well-educated citizens who had made the scientific turn, was clearly a part of the story.

Yet there are other salient patterns in Rosen's research. The 'modern circumstances' to which eugenics enthusiasts compelled their brethren to conform were the 'modern circumstances' of a significantly growing and noticeably changing populace a populace that seemed out of ecclesial and civic reach. During a time when various 'helping' professions were on the rise among middle-class Protestants, there were regions, neighbourhoods and families who seemed out of 'charitable' control. Those who would not be assimilated into the organizational plans of progressives seemed not only extraneous but dangerously chaotic. One facet of the popular appeal of eugenics was its tidy promise to justify those margins. The right algorithms and tools helped those called to tend the boundaries of civic life to conceive of and perform their tasks. Variously unfit people played also, simultaneously, the necessary role of 'the problem'.

The sense that those gifted to do so should take up their civic duty to form a more perfect union was also pervasive. For many mainline Protestants, the call to be a 'good citizen' was tied up with the active formation of civic order. This perhaps puts a different spin on Rosen's well-drawn conclusion. The Protestants most accustomed to their role as well-educated *citizens* had the fewest theological resources to resist the messages of eugenics. The oldest and most unquestionably *American* of the Protestant churches were the first to jump on the eugenics bandwagon.

Is Christianity dysgenic?

An active part of the eugenics conversation was the concern that Christianity itself was dysgenic, inasmuch as charitable giving took from the presumably productive and gave to the presumably parasitic. For those directly serving food and clothing, whether informally or professionally, the question was a practical one. To quote again the Revd Oscar McCulloch (who served many free hot meals through his Congregationalist parish in Indianapolis), the 'benevolent public' insisted on merely 'encouraging' those who lived an 'idle, wandering' existence. For those who were asked indirectly to give, the question was differently practical. They had worked hard for their money, and it was part of Christian stewardship to be responsible givers. A pithy piece in *Eugenics: A Journal of Race Betterment* addressed the question head-on. Asking, 'Is Christianity dysgenic?' and 'Is Christian morality harmful? Over-charitable to the unfit?' the 1928 report included responses from one rabbi and three clergymen serving in New York City.[19] The answers of the three Christian clerics warrant careful reading, for each explicates differently the relationship between traditional faith and science to draw conclusions about Christianity and eugenics.

'Evolution is a term that applies to religion as well as biology', explained the Revd Dr Karl Reiland of St George's Protestant Episcopal Church. The 'early Christian concept that the world should be despised' is 'foot-binding', a 'drag on the progress of religious thought' that 'keeps the church from "stepping out"'. Those who were willing to embrace an aptly evolved religion would recognize that 'the first and foremost salvation of man individually, collectively and universally is the here and now salvation of a healthy heritage'. As Reiland read the relation between 'science and religion', the more conventional form of salvation – of one's soul and body through a saviour – is dependent on the securing of 'sound, safe and sane human beings'; indeed, 'the more we get of these salvations the more likely is any other, and the surer is the kind of religion that can help mankind'. Reiland understood his challenge as a clerical spokesman for eugenics – 'with inexorable certainty of perspective' – to be threefold: 'to revive and accelerate progress along the higher levels of thinking ... to convince the religious conscience that whatever our creed, we are dealing with nature for the fundamental welfare of human nature; and lastly, to be prepared to discover that God was the God of biology before the Bible came on the scene'.

Reiland did not directly answer the immediate, practical question posed by the journal. He went well beyond such service, defining religion within the purview of human evolution and social utility. Situated in this way, Christianity could be cleared of the charge of practising dysgenics. But what is more, Christianity, 'whatever our creed', could be squarely in service to the eugenic aim 'to produce sound, safe and sane human beings'. Evolution applied to religion necessitated boldly taking one's place as a vanguard, eschewing when necessary both traditional doctrine and institutional polity in order to lead. As a final gesture of obeisance to the sponsors of the journal, Reiland suggested that evolved Christians concede the primary, revelatory power of biology.

It is in the implied conversation between Ward and Ryan that Reiland's rather elastic theology hit the road. Harry F. Ward was professor of Christian ethics at Union Theological Seminary (1918–41) and was also a founder of the Methodist Federation for Social Service (1907). In 1928 John A. Ryan was a professor of both political science and moral theology at the Catholic University of America and director of the National Catholic Welfare Council's Social Action Department. Their roles in their respective denominations were parallel, each serving as a professor at his church's premier institution of higher education and as a national spokesman for the progressive wing of his church. Yet their answers were strikingly different.

Ward argued that 'the principles of Jesus' would not necessarily 'weaken and destroy society' unless one was overly 'shortsighted' in the interpretation of those principles. The true answer to the question posed by the journal would not overlook the 'vital fact' that 'the principles of Jesus' call for the 'transformation' of the weak. 'In seeking this goal', Ward argued, Christians with a properly broad vision would accept 'the challenge of removing the causes that produce the weak, including the hereditary factor'. And here Ward was blunt. The 'aim' of proper Christianity is 'a healthy society where all are strong'. The faithful are thus not only allowed but 'compelled' to be 'eugenic'. A 'social ethic based on the principles of Jesus' no less than requires 'the elimination of the weak, not their perpetuation, and this it accomplishes by making them strong and by preventing their production, through both breeding and environment'. By Ward's estimation, Christians had a crucial role to play in a 'coordinated world-wide effort to control population' and 'to the attainment of the highest standards of health and development by all the population'. While Reiland suggested that Christianity should be involved in securing 'sound, safe and sane human beings', Ward cut to the chase. The true goal of Christianity, as Ward read it, is to perpetuate strength and eliminate weakness. That the elimination of weakness would involve the sterilization of the weak was but part of a larger project of forming a 'healthy society'. With this trickled-down brew of Hegel, Darwin and American progressivism, Ward defended Christianity of the charge that it coddles the weak. Between Reiland and Ward, the answer was clear: Christianity would not prevent the social body from vaccinating itself against the unsafe, the unsound and the insane. Quite the contrary – the Christianity of their day was to participate in the process of inoculation.

John Ryan is the figure who most clearly complicates Christine Rosen's suggestion that progressivism and eugenics were inextricably linked. Here was a bona fide progressive, a tireless advocate for the working class and the unemployed poor who helped move his church to heed a radical strand of social thought all but buried in the nineteenth century. Yet Ryan answered the question of the relation between eugenics and Christianity with two points that cut to the heart of the eugenic presumption. First, he reminded his interlocutors that 'society, apart from the human beings composing it, is a mere abstraction' – that is, the social body exists only in and through real, embodied human beings. 'Therefore', Ryan patiently prodded, 'to subordinate the weaker groups to the welfare of society means simply that some human beings are to be made instruments to the welfare of other human beings.' This is all well and good, Ryan warned, for 'one who believes that morality is identical with physical force'. However, 'one who does not identify right with might' will be unable to make an argument for 'treating the weak as of less intrinsic worth than the strong, even though the former may be in the minority'. Ryan left unspoken that the 'one' who identified right with might and considered the weak to be of less intrinsic worth could hardly call him- or herself a Christian.

Then, with a rhetorical twist, Ryan reminded the eugenic readers, and presumably his fellow clergy, that they would do well to consider another, more practical problem. Reiland, Ward and other like-minded Christians might hope to find a hearing now by proving their solidarity with the strong. But if the eugenic programme succeeded, they might eventually find themselves on the receiving end:

> The practical argument against this theory is that once society decides that the weak may rightfully be left to perish, it will extend the principle to all of the so-called inferior classes, so that in the end the 'welfare of society'

will come to mean the welfare of a few supermen, namely those who have been powerful enough to get themselves accepted at their own valuation.

As Ryan narrated the movement, eugenics was primarily about the power of the currently strong to use the apparently weak for the purposes of securing something as nebulous as a 'higher average welfare'. Faced with the temptation of siding with the worthy against the unworthy, the other two clergy had succumbed. Sacrifice your faith, Ryan warned, and tomorrow you may find yourself counted among the weak.

Purge the dross

Many sermons inspired by the AES sermon contests reveal a tragically shortsighted tendency on the part of Protestant clergy to align with the strong against the vulnerable. Three sermons published during the same year (1929) in three different journals by men representing three different denominations may serve as characteristic examples. Edwin Bishop preached 'Eugenics and the Church' at Plymouth Congregational Church in Lansing, Michigan; like the dysgenic/eugenic debate above, the sermon appeared in the official ERO publication from New Haven, *Eugenics: A Journal of Race Betterment*.[20] A second, 'The Refiner's Fire', appeared in the *Homiletic Review*. The Revd Phillips Endecott Osgood, rector of St Mark's Unitarian Church in Minneapolis, had won the 1926 AES eugenics sermon contest with his rousing demand to purge 'the dross' of humanity through eugenics, and the publishers of the ecumenical journal not only found the sermon worthy of print but endorsed the AES contest, declaring with editorial authority: 'Eugenics is an approved thing'.[21] The *Methodist Review* chose 'Eugenics: A Lay Sermon' by George Huntingdon Donaldson, published also in 1929.[22] The three sermons suggest that the popularizing efforts of the scientific eugenicists were quite effective. From an ERO journal to an ecumenical Protestant review to a focused, denominational publication, the message was taking hold. A Congregationalist in Michigan, a Unitarian in Minnesota and a Methodist in New York City each took to the pulpit to affirm (quoting Donaldson) that 'the strongest and best are selected for propagating the likeness of God and carrying on his work of improving the race'.[23]

In an ironic use of anti-Catholic dialect, the Congregationalist in the group noted that the science of heredity was confirming 'Irish Pat's sage dictum that "a family tree is a foine thing if it ben't too shady"'. Science had proven that everything from 'night blindness' to 'a tendency to health and longevity' was 'heritable', and, according to Edwin Bishop, it was through these scientific advances that God called humans to 'participate with him in *conscious* evolution' (emphasis original). Appealing, as did so many eugenicists, to old-fashioned animal husbandry, Bishop argued that 'if we used as much intelligence in human mating as we use in breeding horses and cows' we could prevent 'ills' and encourage 'excellencies'. Would not Jesus himself, Bishop asked, encourage 'any program that would aid children to be physically wellborn?' Again, in a characteristic eugenic move, Bishop referred to the rock-solid proof of numbers. A quantitative study had shown that 'of 476 children born to 144 marriages among feeble-minded folk only six were normal', indicating that 'native ability furnishes the bulk of the basis of achievement'. Those who follow Jesus needed therefore to see eugenics as 'a potential ally' in the holy pursuit of 'racial self-fulfilment'. 'Through neglect of eugenic knowledge and practice', Bishop warned, 'tares are sprouting widely through the wheat.' By forming a prudent alliance with eugenics organizations, Christians could reverse the growing imbalance of the 'well-born' and the 'less favourably born' a menacing imbalance proven by 'carefully assembled data' (probably provided by the AES in their letter of invitation to the sermon contest).[24]

It is for very good reason that Christine Rosen begins *Preaching Eugenics* with the example of Phillips Osgood's 'The Refiner's Fire'. His call, with which we began this chapter, to prove one's 'bowing acquaintance' with the 'larger truths' of eugenics was a rhetorically masterful use of his homiletic gifts.[25] Opening with Malachi 3.3, 'He shall sit as a refiner and purifier of silver', Osgood called God the 'Refiner of the generations', for whom Christians should 'count themselves the agents of his purposes'. Lest any of his hearers or readers ask inconvenient questions about Jesus and the meek or poor, Osgood reminded them: 'Jesus sometimes said ruthless things' if 'men deserved them'. Such was the time again. Jesus 'was superlatively concerned to better the qualities of human living' and testified that 'grapes cannot be gathered from thorns nor figs from thistles'. Citing children's 'inalienable right to life more abundant', Christians were to take up 'the refining responsibility', recognize that 'the future is in our hands', and secure by 'creative forethought' the purity of future generations. Loath to limit himself to Scripture, the Unitarian clergyman reminded his hearers that 'Xenophon, long ago, recommended that slaves should be allowed the reward of children for good conduct'. 'The recommendation', Osgood suggested, 'has merit also for those not slaves.'[26]

From sterilizing 'the criminal' to stigmatizing the 'victim of inheritable malady', Osgood argued with exegetical flourish for the basic tenet of eugenics – the excellent must prevent the propagation of the reprehensible and the pitiable. The present generation had a responsibility to act as 'redemptive helper of the next generation'. If one but compared the Jukes family to the family of Jonathan Edwards (a move first made by A.E. Winship)[27] one might see that heredity is the 'major factor' in 'our cooperation with the Refiner's work'.[28] Here Osgood played on the metallurgy metaphor in two ways. In order to determine who among the Refiner's creatures were called to do the cooperative refining, one needed to look primarily at heritage. In order to determine the proper tools to be used by the refined, one needed only to look at the heredity studies. Osgood's conclusion is horrifically clear:

> God will provide his Spirit to our children's children; why handicap its incarnation? It will be the finer in its manifesting if it need not labor under handicap. The kingdom of God on earth is not an end of growth, but the

beginning of true destiny. Until sin and weakness and disease and pain are done away, we are only starting to commence to get ready to enter into life as it may be. Until the impurities of dross and alloy are purified out of our silver it can not be taken in the hands of the craftsman for whom the refining was done. God the Refiner we know: do we yet dream of the skill or the beauty of purpose of God the Craftsman with his once purified silver? May the time soon come when in refined humanity he can see his own face, clear and unsullied.[29]

Invoking no less than 'the name of God who is Love', Osgood suggested that the culmination of God's creation was dependent on the elimination of suffering. This involved not acts of mercy toward the sick and the poor but acts to secure a future free of those who would 'handicap [the Spirit's] incarnation'.[30] Twice referring to the '*saecula saeculorum*', Osgood places the 'refining responsibility' within the Latin liturgy. The 'forever and ever', 'the ages of ages', becomes, in his sermon, a future dependent on eugenic resolve.[31]

From the husbandman to the blacksmith and, now, to the gardener: George Huntington Donaldson, in his sermon for the *Methodist Review*, used a botanical metaphor to extol the virtues of Christian eugenics. He urged Methodists and fellow humanists to see the 'beautiful and efficient answers' that emerge when one envisions God as akin to the nation's most beloved gardener, California plant-breeder and eugenicist Luther Burbank. 'When we contemplate this patient toiler in his wonderful garden, in fellowship with and conformity to that trinity of creative laws, namely: *heredity, variation*, and *selection*', we may understand 'the progressive creation of better life on this earth.' By reading Genesis with eugenic science in view, one can see that 'the whole creative process described [in the creation story] had been progressive'. Using Unitarian poet William Herbert Carruth's line 'Some call it Evolution, / And [others] call it God', Donaldson himself waxed poetic on the 'third law' of creation. Through selection, God continued to use 'the strongest and best' for 'propagating the likeness of God and carrying on his work of improving the race'. If one reads Scripture carefully, one may detect a pattern: from the story of Joseph's rise over his 'mongrel' brothers to the desert wandering (which 'purged' the weak) to the 'choicest souls' who followed Ezra and Nehemiah, 'those who have been purified' are able to 'see God'.[32]

Exegetes, sociologists, biologists and theologians heartily agreed on this point: 'there is but one fixed and unchangeable thing, and that is heredity'. To bring this point home to his congregation, the preacher sounded a liturgically resonant theme. For Methodists, hymnody was (and remains) a significant part of weekly worship. Donaldson's use of language from Matthew 7.24–25 played effectively upon echoes with the beloved Methodist hymn by Edward Mote, 'My Hope is Built'. While the hymn intones, 'My hope is built on nothing less than Jesus' blood and righteousness', Donaldson proclaimed, 'Heredity is a rock on which we can build with unfailing certainty'. By setting the initial creation within the story of evolution, Donaldson was able to set the new creation through Jesus within the larger story of 'improving the race'. Those who were baptized into Christ became partners in the work of selecting, yanking out and cultivating. With 'proper selections and combinations all good can be produced and all evil eliminated'; the one following Jesus needed to cut down the 'crop of defectives' who 'weaken and burden the race'.[33]

Donaldson showed a characteristic familiarity with the big names of scientific eugenics. By alluding to the founder of British eugenics, Francis Galton, and quoting Henry Fairfield Osborn, president of the American Museum of Natural History in New York (1908–33) and of the Second Eugenics Congress (in 1921), he gave evidence of his own knowledge of the field as well as obeisance to those who were developing the proper gardening tools. His sermon came back to holy horticulture, employing the Johannine Jesus's imagery of vine and branches to link God the Father with God the Selective Gardener:

> So, returning to our Bible account, we read the words of Jesus, 'I am the vine, ye are the branches, and my Father is the gardener,' or husbandman. And just as at the beginning we saw Luther Burbank watching over his gardens, selecting here a fine strain, there another, and taking the pollen of one to unite with the egg cell of the other and so produce a finer fruit or flower, so in this wonderful book, God as Gardener has been watching over humanity.[34]

As the rousing conclusion of his sermon, Donaldson asked whether Christians would have the courage to 'Make democracy safe for the world' by ensuring 'progressive betterment'. In the hymn on which Donaldson relied for a potent echo, it was Jesus's blood that secured the future. Layering metaphor on metaphor, Donaldson suggested that those made 'perfect' by Jesus's 'death upon the cross' were called to make democracy safe by ensuring the 'pure and undefiled' transmission of human blood 'to the coming generations'.[35] The Christian hope is thus *rebuilt* on nothing less than Jesus's blood and Galton's best.

The subtitle for this section, 'Interpreting the historic faith in a modern world', plays on the current motto of the *Quarterly Review*, the United Methodist journal of 'theological resources for ministry'. Rather than a 'modern world', the journal aims to interpret 'the historic faith' in a 'postmodern' one. The editors explain that the journal will 'forthrightly engage the challenges of ministry by bringing the resources of the Christian faith in mutually critical conversation with the issues of our present reality'. In this quest, the articles emerge from 'the context of a distinctively United Methodist and Wesleyan perspective – without ever becoming parochial or narrow-minded'.[36] This promise – to avoid parochialism and narrow-minded thought – seems to most mainline and liberal Protestants today to be a key ingredient of truly *relevant* theology. The editors of the *Methodist Review* issue of 1929 may very well have chosen Donaldson's sermon using primarily that criterion. Donaldson more than amply proved his allegiance to what the editors of the *Homiletic Review* called the 'approved thing' of eugenics. Yet the editors of the current *Quarterly Review* also signal that essays 'interpreting the historic faith in a postmodern world' should emerge from within the Wesleyan tradition. Presumably, the call to remain relevant, to attempt to eschew parochialism or narrow-minded thought, does not trump the call to remain within the range of Wesleyan thought represented by the various Methodist traditions. Striking in the sermons above is the facility with which each of the preachers was able to bend Scripture to suit the eugenic project. Many mainline Protestants had come to believe that, as the pro-eugenics editor of the *Methodist Quarterly Review* put it, because 'the Bible nowhere undertakes to give a detailed account of the process of creation ... it leaves ample room for any theory to which careful scientific investigation may lead'.[37] The scriptural story of salvation in the Old Testament all too swiftly became a story of God's refining, purifying and selecting in order to produce a stronger, heartier stock of humans. Jesus's parables regarding the kingdom of God swiftly became parables for the eugenic separation of human wheat from human chaff. Both Osgood and Donaldson also employed liturgical cues, setting the practices of sterilization and selective mating within the context of worship.

All three preachers were able to interpret 'the historic faith' in a 'modern' (i.e. Darwinian) way: defining Christians as cooperative agents in bringing to fruition God's purposes, which primarily meant the strengthening of the assumedly dominant race in the assumedly Promised Land of the United States. Having gone the way of the modernists in accepting evolutionary science, the mainline denominations represented in these published sermons were able to distinguish themselves from the 'backward' Christian creationists. In the words of Gilbert T. Rowe, editor of the *Methodist Quarterly Review* (1921–28), mainline readers were particularly interested when they found a 'thoroughly Christian' writer who could prove himself 'no hard and fast dogmatist'.[38] One who could so 'thoroughly' narrate the tradition in a way consonant with evolution was a particularly reliable guide to the present and future. But in the process of re-narration, these preachers, and the theologians upon whom they drew, arguably left few theological barriers to the all-encompassing narrative of eugenics. And having accepted as their primary duty that of civic leadership, they left few theological barriers to the racist nationalism of eugenics. Determined to think in modern, patriotic and well-educated ways about the role of the faithful in America, these preachers attained their sophistication at the expense of the vulnerable. They used the lives of others in order to establish their own strength.

Safeguarding the future

Another way to detect the theological moves made by mainline churches during the eugenic era is to read the distillation of eugenic thought in journals aimed at clergy and learned laity. Journals such as the *Methodist Quarterly Review* and *Religion in Life* sought to offer their readers essays relevant for the church's engagement with theological as well as secular disciplines. The volumes of journals from the 1920s and 30s reveal significant interest in the application of evolutionary thought to Christian theology. One *Methodist Quarterly Review* piece, a review essay by Rowe, is characteristic. Rowe sorted through the implications of three books from three different authors, attempting to forge a pathway *To Christ through Evolution* and to make a new theological language that combined *Evolution and Redemption*.[39] A *Religion in Life* piece considered *The Doctrine of Redemption in the Light of Modern Knowledge*, and yet another, by the author of *Do the Ten Commandments Stand Today?* and *Evolution for Christians*, asked, 'Has the concept of humanity a scientific basis?'[40] The journals served to give a particularly Christian (and particularly mainline Protestant) interpretation of the biology and sociology filtering through the university classrooms and over the radio airwaves. Mainline clergy and laity across the country faced the task of thinking at the intersection of evolutionary biology and practical theology. Their efforts are important for understanding the sense that 'Eugenics is an approved thing', again to quote the editors of the *Homiletic Review*.

One local pastor in Missouri, a Revd C.L. Dorris, wrote an extensive essay inspired by Philip Archibald Parsons's *An Introduction to Modern Social Problems* for the 1926 *Methodist Quarterly Review*.[41] This young pastor, trained at Central (Methodist) College and the University of Missouri, was serving in the Methodist Episcopal Church South in the small town of Milan, Missouri, when he felt compelled to write of 'The Impending Disaster', adding 'his voice to the voices in the wilderness warning that unless something is done disaster will soon befall us'. This, the last line of the essay, indicates the exigency of the effort. The sense of moral urgency, of peril and promise, registered even in a town of around 2,000 people, several days' travel from the nearest city.[42] The records of Dorris's training and travels do not indicate that he came to his conclusions after encountering the throngs of immigrants on either shore. Rather, they suggest that the message of eugenics came to him perhaps through his reading at the public library or through a teacher in his congregation. He, in turn, published an essay to be read by local Methodist ministers eager to bring the latest in sociological discernment to their parishes. In it, the rural pastor took in a considerable amount of intellectual territory, borrowing clout from Englishman Henry Havelock Ellis (author of the six-volume *Studies in the Psychology of Sex*), Professor James Quayle Dealey (tenth president of the American Sociological Society), the Reverend Josiah Strong (prominent Congregationalist pastor and author of the anti-urban tract *Our Country*), professor and sociologist Charles A. Ellwood and Harvard president Charles W. Eliot. The essay reads as a pithy, authoritative call to attention and action.

Beginning with the sin of sloth, Dorris admonished readers that 'one of the gravest dangers' among Methodist congregants was 'a lack of pride in providing capable offspring for future generations'. Clearly, there were 'too many physically, mentally, and morally defective' gaining way in society, and Christians needed to relinquish 'the individualistic theory of marriage for personal pleasure'. Quite the contrary marriages needed to be planned with 'racial consequences' in view. Without such prudence, Dorris warned, 'we are going to continue to produce a crop of defectives'. Here Dorris returned to a practical strategy commonly enjoined in eugenic texts of the time – the sacrifice of romantic sentiment for the sake of society. Quoting Ellis's *Psychology of Sex*, Dorris affirmed that 'the birth of a child is a social act' and that the 'community', in being 'invited to receive a new citizen', is 'entitled to demand that that citizen shall be worthy of a place in its midst'. Dorris repeated this crucial point: 'We should demand that each child born is worthy a place in our midst'. The demand required the full arsenal of 'public sentiment in favour of safeguarding the future', for 'public sentiment is one of the most powerful weapons of defence'. In this way, Dorris defined the primary task of his clerical and lay readers: to bring the 'American people' to 'see the dangers threatening us', so that 'they will demand the raising of proper safeguards'. A local pastor from rural Missouri gave an effective populist appeal for the work of eugenics, bringing the civic role of 'safeguarding the future' down from New Haven and Cold Spring and Boston and Manhattan to the grassroots of the heartland. Methodists were to do their considerable part to enforce proper marital standards based on 'the law of the survival of the fittest'.[43]

There were multiple means for averting 'the disaster', ranging from the legal and institutional, to address the inferior, to the more nuanced propaganda necessary to encourage the 'truly better elements of society' to see their civic duty to replenish 'the stock'.[44] Regarding the former, there were laws 'to prohibit the marriage of the unfit', intended to 'eliminate the weaker stock' and 'build up the race through its stronger elements.' There was also the prospect of widespread sterilization. But Dorris argued that sterilization would be insufficient for the 'hopeless types of defectives'. For them, the public should seek means for 'permanent segregation' that would force these otherwise 'expensive' individuals 'to support themselves in properly conducted institutions and colonies'. Yet even with these concerted measures, leaders needed to be vigilant to encourage the reproduction of those representing 'the higher forms of life'. Quoting sociologist James Dealey, Dorris explained that only 'when [society] frowns alike on the large family of the poor and the childless family of leisure' will 'rapid advance' ensue. The 'future civilization' depended as much on the breeding of the 'leadership' class as it did on the institutionalization of the 'defectives'. And while he was on the topic, Dorris reminded the 'leisure class' not to 'confide their children to the care of ignorant and incapable servants'.[45]

Again, this from a local pastor in Milan, Missouri. It is a testimony to the brilliance of those in the eugenics movement that they were able to direct the mainline Protestant desire for class superiority toward the popularization of such notions. Dorris's essay thus reveals a crucial part of the tale. Recall the AES Fitter Family poster warning that a mere 4 per cent of Americans were born with the 'ability to do creative work and be fit for leadership'. The fact that the AES could display this sign at county fairs in Missouri, Michigan and Minnesota without engendering moral outrage is at first glance unfathomable. While it makes some intuitive sense that a Charles Davenport could appeal to the vanity of railroad tycoon-widow Mrs E.H. Harriman with such 'AES calculations', the fact that the hoi polloi in places such as rural Missouri found themselves in the 4 percent rather than the 96 per cent begs for explanation. By writing of the 'impending disaster' from his unassuming Methodist parsonage, Dorris left behind evidence – a vital clue to the eugenics puzzle. He assumed that he and the people who read the *Methodist Quarterly Review* had as central a role to play in civic leadership as did Mr and Mrs Rockefeller. Quoting Professor Charles A. Ellwood, Dorris explained that 'the growing complexity of social life, as social evolution advances, calls for an ever-increasing means of control over individual character and conduct'. As good Christian citizens, Methodists needed to address 'the woes of the world' with 'religion of the right kind' – a religion with proven influence 'in elevating character, in diffusing peace and good will, in fitting men to labour and to endure', and, indeed, 'in lifting mankind to a higher sphere morally and spiritually'. It was a crucial time for the country, with aspirations as high as the sense of economic and demographic vulnerability. With progress, Dorris warned, also comes 'degeneration', and the mainliners of the heartland had a role to play in safeguarding against it.[46] And here Dorris was arguably dead-on. Religious and secular historians alike could have relegated the AES and ERO and the American Breeders Association to mad-scientist status had those organizations not made such headway on everything from state sterilization laws to the popularization of eugenic aspirations among the middle class. The movement *won* with the considerable aid of men such as Dorris and the women and men who heard his call to '*demand that each child born is worthy a place in our midst*'.[47]

The future of the disabled

The residual logic of taint and purity underwrote much that passed for normative parenthood in the twentieth century. Dorris's cold assessment had arguably won the day by the time of the Second World War. The responsible citizenry had a right to demand that each child born was 'worthy a place in our midst'. The turn from overt, coercive eugenics to implicit, voluntary eugenics may be less a sign of the failure of eugenic ideology than a sign of its success. The 'modern', Darwinian, sense of a division between the 'highest' and the 'lowest' in humanity lent scientific legitimacy to the fears at play in the middle-class neighbourhoods that began to flourish in the postwar period. There were some children meant to flourish and others whose lives were insufficiently ordered and wholesome. Distinguishing the one group from the other was a whole set of signals from the shoes on a little boy's feet to the number of braids in a daughter's hair, from the marks on a second-grader's report-card to, eventually, the APGAR score assigned to his newborn sister. This is not even to mention the clear, blatant and often deadly markers of race. The same Methodists Dorris called upon to address the 'impending disaster' of America's degeneration were ready to lead again by way of voluntary eugenics. Promoting the medical tools of 'responsible parenthood', mainline Protestants endorsed family planning and orderly hygiene as practices integral to civic duty. Taking their cues from *Ladies' Home Journal* ('The Magazine Women Believe In'), mainline Protestant mothers employed Ivory Soap and the principles of discriminating reproduction to make sure that their offspring were legitimate inheritors of the Promised Land.

In *The Future of the Disabled in a Liberal Society*, Hans Reinders, Willem van den Bergh Professor of Ethics and Mental Disability at Vrije University in Amsterdam, suggests that the future looks considerably less excellent for those who do not follow the expert advice of such counsellors. Reinders was asked in 1996 to write on genetics and disability for the Dutch Association of Bioethics. His book is an extended moral reflection on the subject, and it directly counters the usual argument that the overriding ethical concern attending prenatal testing is that of simple distributive justice. Indeed, Reinders suggests, the ever-widening distribution of such technologies may in fact weaken the already tenuous commitment of liberal nations to funding disability services:

> Assuming that disabled people will always be among us, that the proliferation of genetic testing will strengthen the perception that the prevention of disability is a matter of responsible reproductive behavior, and that society is therefore entitled to hold people personally responsible for having a disabled child, it is not unlikely that political support for the provision of their special needs will erode.[48]

According to Reinders, the question of civic and social hospitality is key, but political liberalism is not ultimately capable of engendering and fostering hospitality toward people with overt, recalcitrant needs. The norms encircling the liberal axis of individual autonomy cannot easily accommodate lives dedicated to the care of perpetually dependent individuals, or admit the intrinsic value of these individuals. Meticulously considering the policy implications of this tension, Reinders concludes that it is neither within the liberal purview nor within the limits of the practical to address it through legal restrictions on procreative technology and abortion. The predicament facing liberal society, then, is 'cultural', not 'political'.

> The benefits bestowed by love and friendship are consequential rather than conditional, which explains why human life that is constituted by these relationships is appropriately experienced as a gift. A society that accepts responsibility for dependent others such as the mentally disabled will do so because there are sufficient people who accept [this] account as true.[49]

The sense that a life may be rightly mapped on a grid of social use, productivity or beauty begs for an account that can see life otherwise. Mainline Protestants in the United States failed to offer such an account in the past. The efficiently eugenic future now beckons.

Notes

1. This image, with the caption 'The Christian asks: how many; how healthy?', appeared in 1948 in the YMCA/YWCA magazine, the *Intercollegian*, alongside two related articles: Paul Popenoe, 'Surveying the Chances', and Helen F. Southard, 'Planning Parenthood on Campus'. *Intercollegian*, 65 (January 1948), pp. 9–10, 10–11. Popenoe was an active member of the American Eugenics Society (AES) and coauthor of the widely used textbook *Applied Eugenics*. Southard was an advocate for family planning. I am grateful to Rachel Maxson for finding these articles in the midst of her own research and for sharing the reference with me.
2. Phillips E. Osgood, 'The Refiner's Fire', in *Homiletic Review*, 97 (May 1929), pp. 405–409. Christine Rosen gives an account of Osgood's eugenic efforts in her *Preaching Eugenics: Religious Leaders and the American Eugenics Movement* (Oxford: Oxford University Press, 2004). See in particular pp. 3–4, 124–26.
3. Popenoe, 'Surveying the Chances', p. 10.
4. PBS, databank entry, 'Eugenics Movement Reaches its Height (1923)'. A Science Odyssey. http://www.pbs.org/wgbh/aso/databank/entries/dh23eu.html.
5. Watson has gone on record in support of genetic engineering to eliminate physical suffering. He has also grown increasingly blunt in his support of inheritable genetic engineering to avoid people who 'really are stupid' and to make 'all girls pretty'. The relevant sentences, as reported by *The Times*: 'People say it would be terrible if we made all girls pretty. I think it would be great', and 'If you really are stupid, I would call that a disease.' Mark Henderson, 'Let's cure stupidity, says DNA pioneer', *The Times*, 28 February 2003.
6. David Micklos, 'None without Hope: Buck *vs.* Bell at 75', Dolan DNA Learning Center, http://www.dnalc.org/resources/buckvbell.html.
7. *Buck v. Bell*, 274 US at 200, 207 (1927).
8. Micklos, 'None without Hope'.
9. This sign is most likely from an exhibit related to the Eastern States Exposition, 1926. Image from American Eugenics Society Records, courtesy of American Philosophical Society.
10. The language is from the cover story of the 15 March 2004 issue of *Business Week*, featuring a photograph of two parents with their five sons. The article, by Brian Grow, is entitled 'Hispanic Nation', and the cover warns: 'Hispanics are an immigrant group like no other. Their huge numbers are changing old ideas about assimilation. Is America ready?' A graphic that accompanies the story charts 'America's *Bebé* Boom'.
11. The phrase is from the title of a book by William J. Robinson, MD, *Fewer and Better Babies*, published in multiple editions from 1915 to 1938.
12. I have found when presenting this material to non-Southerners that they are shocked to hear that the ERO was based on Long Island, New York, and that the AES had its headquarters in New Haven, Connecticut. The

South has functioned in some ways to provide a second rhetorical chasm, allowing people from elsewhere mentally to dump most American ills in the supposed backwater of Southern culture.
13. See Richard Hofstadter, *Social Darwinism in American Thought* (New York: Braziller, 1959), pp. 29–30; 48.
14. Image from ERO, courtesy of American Philosophical Society.
15. Harry Laughlin, who worked as superintendent of the ERO under Davenport, was the son of a minister.
16. The reference is to Leland Foster Wood, 'The Church and Education for the Family', in *Religion in Life*, 3 (1934), pp. 420–31. Arguing for 'premarital interviews' to promote emotional and social fitness, Wood declared: 'As rapidly as the clergy can be trained for this work, they will enter into a great field of usefulness'. Further: 'While the psychiatrist, the social worker, the judge of the court of domestic relations, the family physician and others render a great service to families, the minister of religion, whose business it is to interpret life as a whole, has his own unique place of service' (ibid., p. 431).
17. Rosen, *Preaching Eugenics*, p. 184.
18. J.S. Haldane, cited in a *Century* report, 'Churchmen and Scientists Discuss Mutual Problems', (25 September 1924), pp. 1244, 1250, 1252. See also J.S. Haldane, *The Sciences and Philosophy* (Garden City, NY: Doubleday, 1929) and Haldane, *The Philosophical Basis of Life* (Garden City, NY: Doubleday, 1931).
19. 'Is Christian Morality Harmful? Over-charitable to the Unfit? Four Religious Leaders Discuss a Charge Sometimes Made', in *Eugenics: A Journal of Race Betterment*, 1 (December 1928), pp. 20–21.
20. Edwin Bishop, 'Eugenics and the Church', in *Eugenics: A Journal of Race Betterment*, 2 (August 1929), pp. 14–19.
21. Osgood, 'Refiner's Fire', pp. 405–406.
22. George Huntington Donaldson, 'Eugenics: A Lay Sermon', in *Methodist Review*, 112 (1929), pp. 59–68.
23. Donaldson, 'Eugenics,' p. 60.
24. Bishop, 'Eugenics and the Church', pp. 16–17.
25. Osgood, 'Refiner's Fire', p. 406.
26. Osgood, 'Refiner's Fire', pp. 405–406.
27. Albert E. Winship, *Jukes – Edwards: A Study in Education and Heredity* (Harrisburg, PA: R.L. Myers, 1900).
28. Osgood, 'Refiner's Fire', pp. 406–407.
29. Osgood, 'Refiner's Fire', p. 409.
30. Osgood, 'Refiner's Fire', pp. 408–409.
31. Osgood, 'Refiner's Fire', pp. 405, 407.
32. Donaldson, 'Eugenics', pp. 59, 60, 63 (emphasis original).
33. Donaldson, 'Eugenics', pp. 63, 65.
34. Donaldson, 'Eugenics', pp. 65–66.
35. Donaldson, 'Eugenics', pp. 67–68.
36. 'About *Quarterly Review*', http://www.quarterlyreview.org/aboutqr.html.

37. Gilbert T. Rowe, 'Christianity and Evolution', in *Methodist Quarterly Review*, 75 (1926), p. 138.
38. Rowe, 'Christianity and Evolution', p. 140.
39. Rowe, 'Christianity and Evolution', pp. 138–41. The essay involves a review of *To Christ through Evolution*, by Professor Louis Matthews Sweet (1925); *Nineteenth-Century Evolution and After*, by Revd Marshall Dawson (1923); and *Evolution and Redemption*, by Revd John Gardner, DD (1925).
40. Ismar J. Peritz, 'Christ and Evolution', review of *The Doctrine of Redemption in the Light of Modern Knowledge*, by George A. Barton, in *Religion in Life*, 4 (1935), pp. 462–64; J. Parton Milum, 'Has the Concept of Humanity a Scientific Basis?', in *Religion in Life*, 5 (1936), pp. 52–63. Milum is also author of *Do the Ten Commandments Stand Today?* (London: Epworth, 1936); *Man and his Meaning* (London: Skeffington, 1945); and *Evolution for Christians* (London: Skeffington, 1933).
41. C.L. Dorris, 'The Impending Disaster', in *Methodist Quarterly Review*, 75 (1926), pp. 720–24, citing Philip Archibald Parsons, *An Introduction to Modern Social Problems* (1924).
42. In 1920 the population of Milan was 2,395; in 1930 it was 2,002. (My thanks to Jason D. Stratman at the Missouri Historical Society for this statistic.)
43. Dorris, 'Impending Disaster', pp. 720–21.
44. Dorris, 'Impending Disaster', pp. 722–23.
45. Dorris, 'Impending Disaster', pp. 721–22.
46. Dorris, 'Impending Disaster', p. 723.
47. Dorris, 'Impending Disaster', p. 720.
48. Hans Reinders, *The Future of the Disabled in Liberal Society: An Ethical Analysis* (Notre Dame, IL: University of Notre Dame Press, 2000), p. 14.
49. Reinders, *Future of the Disabled*, p. 17.

6
Aren't We All Eugenicists Anyway?
Mary B. Mahowald

Introduction

The infamous statement of Justice Oliver Wendell Holmes, 'Three generations of imbeciles are enough',[1] has long been recognized as a shameful example of how eugenics has been practised not only in horrendous situations such as Nazi Germany, but even in a country founded on the principle that 'all Men are created equal'.[2] From the start, the flawed wording of this principle was evident: 'men' was interpreted to exclude women and Negroes. Both groups were excluded from voting rights and other civil rights that white men enjoyed for many years thereafter. The US Supreme Court ruling in *Buck v. Bell* denied another group of people the same basic right that Holmes and his colleagues enjoyed; apparently they did not think that people with mental retardation were 'men'.[3]

Paul Lombardo has made it abundantly clear that the decision in this case was based on an empirically false claim; the three women to whom Holmes referred were not imbeciles at all.[4] Still, even if the claim were true, the decision would still illustrate eugenics, negatively defined as the effort to prevent the birth of 'unfit' individuals.[5] For Holmes, coercive sterilization of a retarded woman was justified in order to ensure that her posterity would not be similarly affected.[6] As Lombardo reminds us, however, Holmes was not a lone champion of this eugenic attitude. American presidents and Nobel laureates alike had been publicly associated with the eugenics movement, which had many supporters in the public-at-large.[7] Apparently, some members of the movement distinguished between practices that were acceptable and those that were not, cautioning that governmental coercion should not be employed in the laudable pursuit of healthy off-spring.[8] But not until the Nazi atrocities demonstrated to the world the horrors to which a eugenic mentality and practice could lead did professional and public support for the movement decline and eventually grow silent.

Advances in genetics and possibilities for manipulating the human genome have resurfaced concerns about eugenics in our day. Typically, these concerns embody the same critique that has been directed against *Buck v. Bell* and against Hitler's atrocities: namely, that they constitute an egregious disvaluing of human beings whose lives and progeny ought to be equally respected. However, one practice that arose between Holmes's and our time has generally escaped concerns about eugenics, despite the similarity between its rationale and that of the Holmes decision. In some quarters this practice has become not only acceptable but expected, leading women who decline it to feel that they are disappointing others, especially their practitioners. I refer to the practice of prenatal testing and termination of affected foetuses.

Prenatal testing is mainly performed to identify foetuses with conditions considered undesirable by parents, practitioners, or society in general. The great majority of these conditions are incurable, although their symptoms or disabling impact may be reducible through treatment and social accommodation. The range of conditions that are identifiable *in utero* has escalated considerably since the human genome has been mapped and sequenced. Chromosomal anomalies and single-gene disorders that affect infants or adults are definitively diagnosable through prenatal testing; well-known examples of these conditions are Down's syndrome, cystic fibrosis, Tay Sachs disease, sickle cell anaemia, and Huntington's chorea. Genetic susceptibility to complex disorders such as breast cancer and Alzheimer disease is also detectable *in utero*, and propensity for some behavioural traits is detectable or likely to be detectable in the near future.[9]

Many foetal disorders are diagnosable through simpler means than genetic tests; these include spina bifida and cleft lip and palate, which are observable *in utero* through ultrasound. However, by far the most common condition for which women are referred for prenatal testing, and for which they seek termination after a positive diagnosis, is Down's syndrome, also called 'trisomy 21'.[10] The rationale that underlies testing and termination for this condition is similar to the rationale of the decision in *Buck v. Bell*: to prevent the birth of a child with mental retardation. However, in contrast to the Holmes decision, which is broadly condemned on legal as well as moral grounds, prenatal testing and termination of a foetus with mental retardation is not only legal, but prevalently viewed as moral. Nonetheless, both practices illustrate the defining intent of negative eugenics: to limit the births of individuals or groups of individuals who are deemed unfit or undesirable.[11]

Ironically, in the years between Holmes's opinion in *Buck* and today, prenatal testing and termination of 'unfit' foetuses have been routinely requested and performed without acknowledging the eugenic nature of these practices. Genetic counsellors, trained to guide their clients to make decisions in conformity with their clients' own values, distinguish between their profession's goals and those of eugenics. Typically, they point to the non-directiveness of genetic counselling and the autonomy of their clients as individuals or couples; eugenics, as they see it, is a coercive practice directed towards whole groups of people.[12] As we will see in what follows, the assumed differences between prenatal termination for Down's syndrome and coercive sterilization of the retarded are not establishable with sufficient clarity to support the claim that the latter, but not the former, constitutes eugenic practice. Even if both practices are eugenic, however, that in itself is not adequate grounds for claiming that they are legally or morally flawed.

As an example of the possible legal and moral acceptability of eugenic practice, consider the behaviour of most pregnant women who intend to bring their pregnancies to term. Most of us who are mothers changed our behaviours considerably during pregnancy, intending thereby to improve the chances of having a healthy child. We took our vitamins faithfully, quit smoking (if we had ever started), avoided aspirin, abstained from ordinary drinks like coffee and Coke as well as alcohol, and, in some cases, endured prolonged bed-rest or hospitalization to avoid premature birth. If positive eugenics is defined as the effort to promote the birth of 'fit' individuals, these behaviours may well be characterized as eugenic. In contrast to forced sterilization of the retarded, however, the efforts of women to do everything they can to have healthy newborns is widely recognized as morally commendable rather than condemnable. Lombardo is right, therefore, to suggest that some eugenic practices are not only morally appropriate but praiseworthy.[13]

Diane Paul has made it clear that the term 'eugenics' can refer to very different kinds of behaviour.[14] Lombardo explores some of these meanings and recounts some of the high and low points in the history of eugenics, concluding with a challenge to find words to substitute for 'imbeciles' in Holmes's infamous statement so that the resultant formulation articulates a sentiment that is morally and socially acceptable.[15] He thus suggests the need for line-drawing, by which we might distinguish between good and bad eugenics.[16] In the first part of this chapter I attempt to do this by approaching the line from both ends: the manifestly bad and the manifestly good expressions of eugenics. My goal is to get as close as possible to where the line should be definitively placed. Preliminarily, however, I examine the broad array of meanings that the term 'eugenics' embraces, and identify the variables that seem to be associated with these different meanings. To the extent that different variables are included in different meanings of eugenics, identifying them helps to clarify what makes some (most?) eugenics bad, some eugenics good, and some eugenics probably neutral.

In the second part of this chapter, I consider the 'disabilities critique' that must be rebutted in support of routine prenatal testing and selective termination of foetuses with disabilities.[17] My analysis suggests a criterion by which to determine whether these procedures constitute good or bad eugenics. Finally (in Part 3), I focus on prenatal testing and termination for Down's syndrome, a condition marked by the same disability attributed to members of the Buck family in the Holmes opinion. Although decisions to avoid having children with Down's syndrome through prenatal testing and termination need not constitute bad eugenics, I argue that broad acceptance of the practice does support the disabilities critique, placing it on the lower end of the spectrum between bad and good eugenics.

1 Eugenics as a spectrum of concepts

Etymologically, the term 'eugenics' comes from the Greek *eugenes*, which means 'well-born'.[18] In light of this derivation, its meaning is as difficult as it has ever been to answer the perennial philosophical question, what is 'the good'? Still, by its literal definition, eugenics does mean something good, not bad: *well*-born, not *ill*-born. Presumably, this meaning is what led some eugenicists of the past to think that the practice they advocated was good, even when others recognized it as good in name but not in fact. Francis Galton, who coined the term in 1883, probably thought he was doing 'good' by championing eugenics as the 'science of improving the stock'.[19] Of course, thinking something is good does not make it so.

To the extent that eugenics is construed as morally objectionable, it is generally associated with coercion. As Paul observes,

> what people object to in eugenics is not the goal, such as improving the health of the population, but the means employed to achieve it. From this standpoint, in the absence of coercion (as reflected in law or obvious forms of social pressure), policies designed with the good of the population in mind are not properly labeled 'eugenic'.[20]

Note, however, that coercion is not an element in the etymology of the term; neither is it included in scientific and dictionary definitions of eugenics as a science by which the human race is improved. Even if the concept or term were mentioned, what constitutes 'coercion' is arguable in its own right. For some, coercion implies the presence of formal, legal barriers to choice; to others, practical impediments such as economic costs and social pressures function coercively.[21] The Holmes decision was coercive in the first sense; in an age in which reproductive freedom is supported by law, women may nonetheless experience coercion in the second sense.[22]

Although I am no more able to define 'the good' definitively than philosophers throughout history have been, I believe it is possible to arrive at an approximate understanding of what constitutes good or bad eugenics by approaching the issue indirectly, starting from the extreme ends of a spectrum of practices that most people consider ethically reprehensible or ethically praiseworthy. Popular approval and prevalent practice do not confer moral validity, which is why the mere fact that prenatal testing and termination after positive diagnosis is widely accepted does not make the practice morally justifiable. Nonetheless, the extreme ends of the spectrum are not just widely endorsed, but universally upheld by reasonable people. This makes the argument for moral validity much more compelling than it would be if controversy prevailed regarding their moral or legal status.

Let us consider, therefore, some examples of activities undertaken or omitted in the name of eugenics that seem manifestly wrong, and some that seem manifestly right or good. On one side, put the genocide committed by the Nazis or other groups who kill classes of people whom they consider undesirable; on the other side, put the health-promoting behaviour of the great majority of pregnant women. Between these opposite ends of the spectrum are a range of behaviours that may be construed as eugenic – sometimes separately, and sometimes in combination;[23] they all fulfil in some way the literal meaning of eugenic as well-born. Many decisions about fertility, whether it is curtailment through contraception, sterilization or abortion, or enhancement through various reproductive technologies, fall within the spectrum of eugenics; so do social policies, laws and cultural norms that affect such decisions. Perinatal decisions may also be eugenic – if their goal is to promote well-bornness.[24]

Prenatal testing and selective abortion are at neither end of the spectrum between good and bad eugenics. By broad social agreement, the *Buck v. Bell* decision belongs closer to the bad end.[25] However, determination of where a particular behaviour belongs on the spectrum depends on multiple variables, some of which are identifiable through examination of the practices that are clearly locatable at either end of the spectrum. The following characteristics distinguish between the two extremes:

Table 6.1 Characteristics of 'good' and 'bad' eugenic practice

Nazi genocide	Health-promoting behaviour during pregnancy
Coercive intervention by state or government	Autonomous decisions by potential parents
Directed at born persons as a group	Directed to potential children as individuals
Terminating their lives	Supporting their lives
To avoid a specific trait or traits	To promote health or other conditions
Judged by state to be undesirable	Judged by potential parents to be desirable

Notice that one side opposes and the other respects the autonomy of those who are directly affected. Note too that one side involves people already born, while the other involves individuals that have not been born and may not even have been conceived. One side is eugenic practice through termination, not just prevention, of already-born individuals who are considered undesirable; on the other side is the avoidance of harms and promotion of benefits to intended offspring. On one side, the practice is driven by the state or government and directed towards an entire group of people who are defined by a single trait or set of traits. On the other side, the practice is driven by individual women or couples and directed towards potential children as individuals.

As Aristotle observed long ago, the good of society generally outweighs the good of the individual as such.[26] Based on that priority, the implicit emphasis on social welfare in the left column is a good, but other characteristics in that column are not. In contrast, the characteristics on the right are generally understood in a positive moral light. Coercion, for example, carries a moral onus that respect for autonomy does not – even though both are sometimes justifiable and sometimes not. And decisions to terminate lives are obviously tougher (and for pacifists, impossible) to justify than decisions to extend life – because life is a prima-facie good.[27] Terminating lives is even tougher to justify when the individuals to be killed are already born, and the sole criteria for termination are single traits or sets of traits found in whole groups of people who may also be killed by those criteria. In contrast, the lives to be supported on the right are seen holistically, as individual potential children whose worth and right to life are not definable solely on grounds of any single trait or sets of traits.

The *Buck v. Bell* decision is on the left side of the eugenics spectrum because it fulfils all but one of the characteristics listed under Nazi genocide. The Supreme Court's ruling in *Buck v. Bell* authorized the forced sterilization, but not the killing, of 'imbeciles'. Nonetheless, it constituted government endorsement of coercive intervention to avoid a specific trait deemed socially undesirable by state legislators. Worse, the Holmes decision purported to effect its eugenic goal by preventing individuals from exercising a right that is central to many people's lives, i.e., the right to have a child.[28] Admittedly, some people with disabilities may be incapable of raising a child or, at least, raising one by themselves. Many are, nonetheless, capable of biological and social parenthood. So the Holmes decision is only as much removed from the far left as sterilization is from homicide. Moreover, as Lombardo makes clear, the assessment of someone as an 'imbecile', or so impaired as to justify sterilization, may be questionable even on empiric grounds.[29]

Prenatal testing is of course separable from termination of affected foetuses. When it is considered separately, prenatal testing may be not only *close to* but *at* the right end of the spectrum of eugenics. Some women seek testing with no intention of terminating their pregnancies if the foetus is found to have an anomaly. They may request tests solely to identify a condition that is potentially and effectively treatable *in utero*, to determine a mode of delivery that is likely to optimize the outcome for the child, or simply to prepare themselves or other family members for the birth of an affected child. In such cases, the testing is either eugenically neutral or 'good eugenics'.

When prenatal testing is undertaken to identify anomalies and terminate affected foetuses, it belongs closer to the left side of the spectrum. Two factors distinguish this from forced sterilization: the eugenic decision is made autonomously by the pregnant woman rather than by government imposition; and the life of the foetus, rather than the capacity for reproduction, is thereby ended. Governmental coercion puts sterilization closer to the far left, but direct killing of the foetus may be just as bad or worse if the foetus is imputed to have moral status. This brings us to the charge levelled by some people with disabilities against those who support prenatal testing and termination of affected foetuses. To them, these routine practices clearly constitute bad eugenics.

2 The disability rights critique of prenatal testing and selective abortion

The link between genetics and advocacy for people with disabilities has precipitated 'the disability rights' critique of prenatal testing and selective abortion, and 'the expressivist argument' with which the critique is associated.[30] According to Erik Parens and Adrienne Asch, the critique involves three main claims.[31] First, prenatal diagnosis undercuts recognition of the extent to which the meaning and impact of 'disability' are socially constructed; second, it implies unwillingness of parents to accept an imperfect child; and third, it usually involves inadequate understanding of the disabilities it attempts to avoid.[32]

Prenatal testing probably does undercut recognition that disabilities are largely socially constructed. Nonetheless, it is possible to support prenatal testing while reducing the disadvantagous impact of its social construction. Positive prenatal diagnosis generally leads to termination, but it is the termination, rather than the diagnosis, that is most problematic from a disability rights perspective. In fact, the diagnosis may be undertaken to facilitate interventions on behalf of the disabled or even to ensure that the intended child is affected with a specific disability.[33] (I will ignore here the fact that some supposed 'disabilities' are not viewed by people with those conditions as disabilities.)

The second claim, that prenatal testing implies parental unwillingness to accept an imperfect child, is not necessarily true; rather, it implies the unwillingness of parents to accept a foetus with certain disabilities if this can be avoided through testing and termination or treatment. What is pivotal here, in part because imperfect newborns (and children) whose parents accept and love them are commonplace, is that the parents who terminate after positive diagnosis do not consider the foetus a child at all, whether perfect or imperfect. At most the foetus is a potential child, and the *potential* of having a child with disabilities is what is avoided. So long as the foetus is not morally comparable to a person who is disabled, testing and termination to ensure that ill-bornness is prevented may be morally equivalent to contraception for better spacing of offspring.

The third claim of the disability rights critique is that prenatal diagnosis usually involves inadequate understanding of the disabilities it attempts to avoid. This claim is true in most cases despite the efforts of genetic counsellors to provide their clients with all of the information relevant to their decisions. It is hardly controversial that women or couples deserve to be maximally informed about the disabilities for which they may be tested. However, knowing more about a condition does not necessarily mean that a decision to terminate is less likely. For at least one condition, Down's syndrome, the opposite seems to be the case. (I will return to this point later.)

The 'expressivist argument' with which the disability rights critique is associated is stronger than the preceding claims.[34] Simply put, the argument is that prenatal testing sends the message to people with disabilities that their lives are not worth living. As Asch observes: 'a single trait stands in for the whole, the trait obliterates the whole' with 'no need to find out about the rest'.[35] However, support for a woman's decision to terminate a foetus assumes the priority of her choice over the life of the foetus, regardless of whether it is disabled. It should be possible, therefore, to support a right to testing and termination without practising the discrimination towards people with disabilities that apparently motivated the *Buck* decision.

Regardless of whether abortion is legal or moral, prenatal testing and selective abortion to avoid the birth of children who are disabled may exemplify bad eugenics. Although decisions to terminate an affected foetus are assumed to be made autonomously by individuals, the rationale for termination is to prevent the birth of a child whose trait, identified as undesirable, 'stands in for the whole'.[36] Occasionally, the rationale for the termination is the best interests of the potential child; in other words, it seems better for a particular foetus not to be born because its inevitable 'ill-bornness' is so severe. Even with very severe anomalies, however, the predominant experience of the child is rarely, if ever, one of suffering unless he or she is not given adequate care after birth.

Provision of 'adequate care after birth' is usually much more demanding and difficult for parents of children with disabilities than for other parents. Few have the resources, whether economic or psychosocial, to meet the challenge alone; yet society often seems to expect them to do so. A healthy woman who lacks the necessary resources for providing adequate care of a healthy infant may terminate her pregnancy solely on grounds of her inability to care or lack of social supports for doing so. While such decisions are morally problematic for various reasons, they do not constitute bad eugenics. If it is not bad eugenics for a woman to choose abortion because of her inability to care for a child who is not disabled, neither is it bad eugenics for her to choose abortion solely because she is unable to care for one who is disabled and no one else is willing or able to provide care. Acknowledgment of one's inability to care for another is not equivalent to rejection of another because of a condition or trait that renders the other unworthy of care. Accordingly, a criterion by which we may determine whether prenatal testing and termination of an affected foetus illustrates 'bad eugenics' on the part of the woman who chooses these procedures is that *the mere fact of the disability is not the pivotal reason for her choice*.[37] Other reasons may be adequate or inadequate in their own right, but they do not constitute the bad eugenics of discrimination against the disabled, nor do they imply that life with disability is not worth living. Other possible reasons for testing and termination are the avoidance of health risks to the pregnant woman and her responsibilities for other children or adults.

If prenatal testing and termination are performed *solely* to avoid the birth of a child with a specific trait, the procedures are closer to the left end of the spectrum between bad and good eugenics. Down's syndrome is a chromosomal anomaly associated with the level of retardation that the *Buck v. Bell* court apparently wanted to avoid in future generations; it thus seems to illustrate this leftward leaning. Because Down's syndrome is tested for so routinely in the prenatal setting, it merits careful scrutiny as a potentially acceptable substitute for 'imbeciles' in the *Buck v. Bell* case. Few people with Down's syndrome are classifiable as 'imbeciles'. Like the Bucks, they may be educated and live satisfying lives despite their mental limitations. When the justices formulated their ruling in the Holmes decision, they did not have the benefit of prenatal tests to determine whether the alleged retardation of Emma, Carrie and Vivian Buck was hereditary; they apparently based their judgement on the (inaccurate) observation that the retardation had occurred in all three generations.[38] Even if the *Buck* court had been correct about the alleged retardation and its hereditary character, it could not have definitively predicted its degree of impact on future generations. Today we can definitively diagnose Down's syndrome and some other anomalies *in utero*; in many cases, however, we cannot definitively predict their impact on affected individuals or on society in general.

3 Prenatal testing and termination for Down's syndrome

Down's syndrome, the most frequently identified cause of mental retardation, occurs in about one in 770 newborns.[39] This incidence is lower than it was prior to the advent of prenatal testing and the availability of selective abortion. However, because the life-span of affected individuals has improved considerably during the past few decades,[40] the actual number of people with Down's syndrome in the general population has been increasing, just as it has with regard to other conditions associated with a shortened lifespan.[41] Although referrals for prenatal testing may be based on general screening tests or positive family history, most referrals are based on maternal age of 35 years or more. The latter rationale stems from the fact that the risk of chromosomal anomalies increases with age, and 35 is the approximate age at which the risk of foetal loss or damage to the foetus through amniocentesis itself is about equal to the risk of having an affected foetus.[42] The actual risk of Down's syndrome in a woman who is 35 is one in 385; the risk of her having a foetus with other anomalies is one in 434, making her total risk of chromosomal anomaly one in 204.[43] The foetal loss-rate for midtrimester amniocentesis is 1 per cent, and for transcervical chorionic villus sampling is 0.5–1 per cent over the general population risk.[44]

In comparison with the symptoms of other prenatally diagnosable anomalies, the common symptoms of Down's syndrome are well known to most people. Most notable is mental retardation, found in all affected persons, but the degree of retardation, ranging from moderate to severe, is not predictable prenatally.[45] Most people are also familiar with facial features associated with Down's syndrome; they are less likely to be aware of medical problems that occur more frequently in those affected (e.g. about 40 per cent have congenital heart disease).[46] Most of these medical problems are as treatable as they would be in other patients.

People with Down's syndrome are described as having 'warm, loving personalities and enjoy[ing] art and music'.[47] Some parents claim they are easier to raise than their unaffected offspring.[48] Because children with Down's syndrome are apparently happy, preventing their birth can hardly be justified as a means of preventing suffering. A more honest rationale is prevention of the burden of their care to their family members or to society. While this rationale may also be morally problematic, it is not equivalent to a claim that their lives are not worth living. Consistent with the criterion I have suggested, so long as the reason for prenatal testing and termination is not the disability as such, testing and termination for Down's syndrome does not belong on the left side of the eugenics spectrum.

While individuals are unable to care adequately for a child in some instances, the same is hardly true for society as a whole, at least in the developed world. Collectively, society has all the resources necessary to care adequately for all of its people: healthy newborns, those with disabilities, or anyone who needs care that is not available through parents or other family members. Accordingly, society in general does not have the justification that some pregnant women may have for testing and abortion of foetuses whose subsequent care may be impossible for them to provide. So why has prenatal testing and termination of affected foetuses, particularly those with Down's syndrome, become so widely accepted by society? One reason is that foetuses do not count as persons under the law of the land.[49] Although some foetuses are developmentally older and healthier than some premature infants, they do not have rights comparable to those of born individuals. As long as a clear line can be drawn at birth, decisions to terminate the developing organism prior to that time are separable from those made after birth, regardless of whether it is well-born or ill-born.

Another reason is that society, through its policy-makers and those who influence public opinion, really does want to reduce the number of people who are mentally retarded in the general population; it may focus on Down's syndrome because its presence is more easily recognizable than other conditions associated with mental retardation. In general, it wants to 'improve the stock' and perhaps avoid the costs of care by eliminating or at least reducing the numbers of a particular group of people by encouraging testing in women and supporting the abortion of foetuses that test positive for Down's syndrome. That this rationale has been effective seems clear from the fact that most women who are told that their foetus has this anomaly choose to terminate their pregnancies more quickly than when they are given other foetal diagnoses, some of which have more devastating medical consequences.[50]

Broad acceptance of testing and termination for Down's syndrome is thus triggered by a society that generally supports the termination of lives considered undesirable because of a specific trait, namely, mental retardation, and possibly because of the appearances that are characteristic of people with Down's syndrome. Although decisions for prenatal testing and termination are usually thought to be autonomous, some individuals report that they feel pressured by physicians and others to undergo prenatal testing and encouraged to terminate when the result is positive.[51] To the extent that this is so, prenatal testing and termination of affected foetuses cannot be considered 'good eugenics'.[52] If the decisions are imposed by others, the principal difference between Nazi eugenics and prenatal testing and termination for a foetus with Down's syndrome is that foetuses are not born persons. Obviously, this is an important distinction, but one that still places it with the *Buck v. Bell* court on the left side of the eugenics spectrum. In other words, with regard to routine testing and termination for this particular anomaly involving mental retardation, the decisions of individuals or couples are eugenically neutral, so long as conditions other than the disability itself form the rationale for the decision to terminate. Depending on the other reasons for the testing and termination, and assumptions about the moral status of the foetus, these decisions may be ethically justified.

Social attitudes and practices regarding prenatal testing for Down's syndrome are another matter. I believe these illustrate bad eugenics for a number of reasons. Principal among these reasons is a deepseated ableism on the part of society's leaders, who, having benefited by the abilities they currently enjoy, rarely recognize that these are mainly a matter of luck or fortune rather than deserved or earned. Even in a culture of political correctness, where attempts to ignore differences are manifest, this ableism prevails. Ironically, it is reinforced by ignoring differences and therefore doing nothing to correct the inequalities associated with them. This ethos of ableism no doubt influences individual women and couples to conform to its standard by avoiding the birth of a child who is disabled.

Conclusion

Recall the expressivist argument with which the disability rights critique is associated: prenatal testing sends the message to people with disabilities that their lives are not worth living. This argument is well-supported, I believe, in the encouragement pregnant women typically receive to undergo prenatal testing when they are 35 years or older. Some women report that they are more than encouraged; they are *expected* to undergo prenatal testing because of the supposedly high risk of a chromosomal anomaly, especially Down's syndrome.[53] To a lesser degree women are expected and encouraged to terminate the pregnancy if the foetus is affected.[54] In other words, the single trait of mental retardation and other traits associated with Down's syndrome stand in for the whole of the potential person, and as Asch puts it, there is no need to find out about the rest of the person because 'the trait obliterates the whole'.[55] In the Holmes decision, the trait of mental retardation obliterated the right of three people to become parents. In prenatal testing and termination for Down's syndrome, the trait of mental retardation obliterates the person that the foetus may become. In this narrow context, then, I propose a substitute for Holmes's infamous statement, one that I consider not only legally supportable but morally demanded: 'Three generations of people with mental retardation are not enough'.

Notes

This chapter was originally published in the *Florida State Law Review*, 30.2 (Winter, 2003), pp. 223–34, as a commentary on Paul Lombardo's 'Taking Genetics Seriously: Three Generations of ??? Are Enough', *Florida State Law Review*, 30.2 (2003), pp. 191–218.

1. *Buck v. Bell*, 274 US at 200, 207 (1927): http://caselawlp.findlaw.com/scripts/getcase.pl?=US&vol-274&invd-200. Hereafter, I shall refer to this decision either as *Buck v. Bell* or as the Holmes decision.
2. US Declaration of Independence (1776), para. 2.
3. See *Buck v. Bell*, 274 US at 207.
4. See Paul A. Lombardo, 'Three Generations, No Imbeciles: New Light on *Buck v. Bell*', *New York University Law Review*, 60 (1985), pp. 30–62.
5. Daniel J. Kevles cites the work of Francis Galton, Karl Pearson and other eugenicists who 'equated fitness with physique and mental ability, and assumed that it was centered in the middle, and particularly the professional, class'. Daniel J. Kevles, *In the Name of Eugenics: Genetics and the Uses of Human Heredity* (New York: Knopf, 1985), p. 32. With Galton's approval, C.W. Saleeby proposed the distinction between negative and positive eugenics. (p. 321). Negative eugenics was 'intended to encourage the socially disadvantaged to breed less – or, better yet, not at all' (p. 85). Positive eugenics 'aimed to foster more prolific breeding among the socially meritorious'.
6. *Buck v. Bell*, 274 US at 207.
7. Paul A. Lombardo, 'Taking Eugenics Seriously: Three Generations of ??? Are Enough?' *Florida State University Law Review*, 30 (2003), pp. 191, 208.
8. Lombardo cites Alexander Graham Bell as an example of those who distinguished between eugenic practices that were acceptable and those that were not. Bell opposed coercive legal measures and advocated efforts to improve undesirable traits rather than eradicate them. Even if the Buck women had actually been mentally retarded, he still would not have agreed with the Holmes decision. (See Lombardo, 'Taking Eugenics Seriously', pp. 211–13.)
9. Behavioural traits related to genetics include tendencies to alcoholism, obesity, sexual orientation, dyslexia, athleticism, and timidity. Mary B. Mahowald, *Genes, Women, Equality* (Oxford: Oxford University Press, 2000), p. 246 (documenting these and other examples of behavioural traits attributed to genetics).
10. 'Trisomy' refers to the fact that the affected person has an extra chromosome, i.e., three chromosomes instead of two; '21' indicates which chromosome pair is affected. See Cathleen M. Harris and Marion S. Verp, 'Prenatal Testing and Interventions', in *Genetics in the Clinic: Clinical, Ethical, and Social Implications for Primary Care*, ed. Mary B. Mahowald (Philadelphia, PA: Mosby 2001), pp. 59, 60; Cynthia Powell, 'The Current

State of Prenatal Genetic Testing in the United States', in *Prenatal Testing and Disability Rights* (ed. Erik Parens and Adrienne Asch; Washington, DC: Georgetown University Press, 2000), pp. 44-45. The condition is called 'Downs' syndrome' after Sir John Langdon Haydon Down, who first described its symptoms, comparing them with those of 'Mongols'. See Rayna Rapp, *Testing Women, Testing the Foetus: The Social Impact of Amniocentesis in America* (London: Routledge, 1999), pp. 295-96. Rapp also cites Down's syndrome as the most common condition for which women seek prenatal testing (p. 223). The actual reason for referral in these cases is 'advanced maternal age', which generally means 35 or older. Although the risk of having a child with Down's syndrome increases with maternal age, most children with Down's are born to younger women, who are not routinely referred for prenatal diagnosis as are older women. Younger women are referred for prenatal tests if they have a family history of a hereditary disease, if they have already had an affected child, or if screening tests suggest a need for definitive testing. See Marion S. Verp, 'Prenatal Diagnosis of Genetic Disorders', in *Principles and Practice of Medical Therapy in Pregnancy* (ed. Gloria E. Sarto; Norwalk, CT: Appleton & Lange, 1992); Glenn Schemmer and Anthony Johnson, 'Genetic Amniocentesis and Chorionic Villus Sampling', *Obstetrics and Gynecology Clinics, North America*, 20 (1993), pp. 497; 515-16.

11. See Kevles, *In the Name of Eugenics*, p. 85 (discussing Saleeby's distinction between negative and positive eugenics).
12. See generally Mary B. Mahowald, *et al.*, 'Genetic Counseling: Clinical and Ethical Challenges', in *Annual Review of Genetics*, 32 (1998), pp. 547, 548-50 (discussing the origin, nature, and goals of genetic counselling).
13. See generally Lombardo, *Taking Eugenics Seriously*.
14. See, for example, her excellent analysis of different meanings of eugenics and their implications in Diane B. Paul, 'Eugenic Anxieties, Social Realities, and Political Choices', in *Are Genes Us? The Social Consequences of the New Genetics* (ed. Carl F. Cranor; New Brunswick, NJ: Rutgers University Press, 1994), pp. 143-45.
15. See generally Lombardo, *Taking Eugenics Seriously*.
16. Lombardo, *Taking Eugenics Seriously*.
17. This critique, also called '[t]he disability[ies] rights critique' is well-developed by Erik Parens and Adrienne Asch in 'The Disability Rights Critique of Prenatal Genetic Testing: Reflections and Recommendations', in *Prenatal Testing and Disability Rights* (Washington, DC: Georgetown University Press, 2000), pp. 12-13.
18. *The Compact Oxford English Dictionary* (Oxford: Oxford University Press, 1993), p. 536.
19. Ruth Hubbard and Elijah Wald, *Exploding the Gene Myth: How Genetic Information is Produced and Manipulated by Scientists, Physicians, Employers, Insurance Companies, Educators, and Law Enforcers* (Boston, MA: Beacon Press, 1993), p. 14 (quoting Francis Galton, *Inquiries into Human Faculty* [London: Macmillan, 1883], pp. 24-25).

20. Paul, 'Eugenic Anxieties', p. 145.
21. Paul illustrates this point with regard to the different political perspectives. A classical liberal or libertarian, she says, would consider the potential parents of a child with Down's syndrome 'free to abort the fetus or bring it to term', whereas an egalitarian liberal or socialist would claim that the '"downstream" costs of caring for a severely handicapped child' may limit their freedom to bring an affected pregnancy to term (Paul, 'Eugenic Anxieties', p. 146).
22. Feminist philosophers have recently formulated a concept of 'relational autonomy', which critiques a narrow or literal conception of freedom on grounds that individuals are not adequately definable atomistically; rather, our ongoing relationships to others are inseparable from our autonomous decisions. See generally Catriona MacKenzie and Natalie Stoljar (eds), *Relational Autonomy: Feminist Perspectives on Autonomy, Agency, and the Social Self* (New York: Oxford University Press, 2000). 'Relational autonomy' also takes into account environmental limitations and social pressures on the decisions of individuals (MacKenzie and Stoljar, *Relational Autonomy*).
23. Prenatal testing, for example, is separable from termination of a foetus, and may in fact be associated with the desire to treat rather than eliminate an affected foetus.
24. Ironically, once an individual is ill-born, regardless of the degree in which well-bornness was pursued before birth, the medical options available are largely anti-eugenic rather than eugenic. By prolonging the lives of those who are not well-born so that they reach reproductive age, medical practitioners facilitate the births of more people who are ill-born. If germ-line gene therapy is ever successful in humans, this anti-eugenic propensity of health care could be reduced; I doubt, however, that even then it would be overcome.
25. As Paul Lombardo observes, '[M]any of the commentaries on *Buck* describe the case as an aberration traceable to the "eugenics craze" of the Progressive Era'. Lombardo, 'Taking Eugenics Seriously', p. 32. Presumably, an 'aberration' of the 'craze' would be even more problematic than the craze itself.
26. See, e.g., Aristotle, 'Politics', in *The Basic Works of Aristotle* (ed. Richard McKeon, trans. Benjamin Jowett; New York: Random House, 1941), p. 1129: '[T]he state is by nature clearly prior to the family and to the individual, since the whole is of necessity prior to the part'.
27. It is *at least* a prima-facie good. Beyond that minimal claim, it may be argued that life is a necessary condition for all other goods attributed to living entities.
28. See Mary B. Mahowald, *Women and Children in Health Care: An Unequal Majority* (New York: Oxford University Press, 1993), pp. 93–97 (discussing the meaning of a 'right', distinctions between legal and moral, positive and negative, absolute and relative rights, and differences between the right to have a child and the right to reproduce).

29. Lombardo, 'Taking Eugenics Seriously', pp. 57–63.
30. For a recent survey of different positions on the morality of abortion and arguments in support of them, see Susan Dwyer and Joel Feinberg (eds), *The Problem of Abortion* (Belmont, CA: Wadsworth, 3rd edn, 1997).
31. Parens and Asch, 'The Disability Rights Critique', pp. 12–13.
32. Parens and Asch, 'The Disability Rights Critique'.
33. Admittedly, requests to ensure disability in children are both legally and morally problematic.
34. Parens and Asch cite Allen E. Buchanan for developing the 'expressivist argument' that they elaborate and critique. Parens and Asch, 'The Disability Rights Critique', pp. 13–17 (citing Allen E. Buchanan, 'Choosing Who Will Be Disabled: Genetic Intervention and the Morality of Inclusion', in *Society, Philosophy and Policy*, 18 (1996), p. 13.
35. Parens and Asch, 'The Disability Rights Critique', p. 13.
36. Parens and Asch, 'The Disability Rights Critique'.
37. I developed the rationale for this criterion in Anita Silvers, David Wasserman and Mary B. Mahowald, *Disability, Difference, Discrimination: Perspectives on Justice in Bioethics and Public Policy* (New York: Rowan & Littlefield, 1998), pp. 236–39.
38. See *Buck v. Bell*.
39. Ricki Lewis, *Human Genetics: Concepts and Applications* (Duberque, IA: William C. Brown, 2nd edn, 1997), p. 212.
40. In the USA, for example, the median age at death of people with Down's syndrome increased from 25 years in 1983 to 49 years in 1997. Quanhe Yang *et al.*, 'Mortality Associated with Down's Syndrome in the USA from 1983 to 1997: A Population-based Study', in *The Lancet*, 23 March, 2002, p. 1019.
41. Cystic fibrosis is another condition for which improvements in treatment have led to increased lifespan in affected individuals. Although the fertility rate of women with cystic fibrosis is less than that of their healthy counterparts, many who survive into their reproductive years have children. In contrast, men with Down's syndrome or with cystic fibrosis are generally infertile, and very few cases of pregnancy in women with Down's syndrome have been reported. Regarding fertility in people with cystic fibrosis, see Robert C. Stern, 'Cystic Fibrosis and the Reproductive Systems', in Pamela B. Davis (ed.), *Cystic Fibrosis* (New York: Macel Dekker, 1993), p. 381. Regarding fertility in people with Down's syndrome, see Paul T. Rogers and Mary Coleman, *Medical Care in Down's Syndrome: A Preventive Medicine Approach* (1992), pp. 196–98.
42. Verp, 'Prenatal Diagnosis of Genetic Disorders', *Obstentrics and Gynecology Annual*, 25.4 (December 1982), pp. 635–58.
43. Verp, 'Prenatal Diagnosis of Genetic Disorders'.
44. Verp, 'Prenatal Diagnosis of Genetic Disorders'. See also Schemmer and Johnson, 'Genetic Amniocentesis and Chorionic Villus Sampling', *Obstentrics and Gynecology Annual*, 20 (1993), pp. 515–16.

45. Moreover, the majority of people with mental disabilities are only mildly retarded. According to Anita Silvers, citing Justice Thurgood Marshall in *City of Cleburne v. Cleburne Living Center*, 473 US 432, pp. 461–66 (1985) http://caselaw.lp.findlaw.com/cgi-bin/getcase.pl?US&vol=473&invol=432 (J. Marshall, concurring in the judgment in part and dissenting in part), over 90 per cent of people labelled with mental retardation would not have been considered disabled in other periods of history, and the capabilities of many in this group are more comparable to those of non-retarded people than to the capabilities of severely retarded individuals (E-mail from Anita Silvers, Professor of Philosophy, San Francisco State University, to author; 10 May, 2002).
46. Other potentially life-threatening disorders with a higher-than-normal incidence in people with Down's syndrome are gastrointestinal disease and leukaemia. Rogers and Coleman, *Medical Care in Down's Syndrome* (New York: Marcel Dekker, 1992), pp. 78–81. One of the gastrointestinal disorders more prevalent in infants with Down's syndrome than in other infants is oesophageal atresia. In 1982 parents in Bloomington, Indiana, refused consent for surgery to correct this life-threatening condition in their newborn with Down's syndrome. Litigation regarding the refusal (which led to the infant's death on the sixth day of life) provoked various efforts of the federal administration and Congress to mandate life-saving treatment in similar circumstances. For a summary of the case and related legislative efforts, see Mahowald, *Women and Children in Health Care*, pp. 170–72, 181–82.
47. Lewis, *Human Genetics*, p. 210. That this description is rather stereotypical should be acknowledged. Gibson observes that people with Down's syndrome are widely imputed to have traits that are contradictory; they are alleged, for example, to be 'affable, mischievous, docile, aggressive, affectionate, stubborn, pleasing and self-willed'. David Gibson, *Down's Syndrome: The Psychology of Mongolism* (New York: Grune & Stratton, 1978), p. 111. Of course, individuals with Down's syndrome vary considerably in their manifestation of these traits, and stereotypes are not necessarily applicable to all members of a class.
48. See generally P. Gunn and P. Berry, 'The Temperament of Down's Syndrome Toddlers and their Siblings', in *Journal of Child Psychology and Psychiatry*, 973 (1985), p. 26; Brian E. Vaughn *et al.*, 'Short-term Longitudinal Study of Maternal Ratings of Temperament in Samples of Children with Down's Syndrome and Children Who Are Developing Normally', *American Journal of Mental Retardation*, 607 (1994), p. 98.
49. The legality of abortion assumes that foetuses do not have rights comparable to those of born individuals. See generally *Roe v. Wade*, 410 US at 113 (1973) (http://biotech.law.lsn.edu/cases/reproduction/roevwade.htm). However, damages and insurance payments are sometimes awarded to pregnant women on grounds that their foetuses are harmed or prevented from being born. See, e.g., *Transamerica Ins. Co. v. Bellefonte Ins. Co.*, 490 F. Supp. 935 (1980).

50. According to Rapp, abortion after a prenatal diagnosis of Down's syndrome is 'almost automatic' because the women whose foetuses are affected are generally familiar with symptoms of the condition. With other diagnoses, they tend to seek more information before making their decisions (Rapp, *Testing Women, Testing the Foetus*, pp. 223–25).
51. Unlike genetic counsellors, the obstetricians who routinely offer and provide prenatal diagnosis to women of 'advanced maternal age' are trained to be directive rather than non-directive with patients. The goal of these physicians is to ensure that the woman and her potential child are both healthy; to many, accomplishing that goal may require testing and termination of an affected foetus. Given the usual power discrepancy between pregnant patient and physician, and the woman's dependence on him or her for care, this attitude entails at least a subtle form of pressure to do what the physician wishes.
52. Presumably, this is why the American Medical Association (AMA) warns against the 'subtle or passive eugenics brought about through a combination of social pressures' to employ existing genetic reproductive technologies' (Council on Ethical and Judicial Affairs, AMA), 'Ethical Issues Related to Prenatal Genetic Testing', *Archives of Family Medicine* (1994), pp. 633–5. The AMA Council acknowledges that these technologies already provide the basis for decisions about the worth of individual lives, and that this 'may constitute an extremely dilute but acceptable form of eugenic selection'.
53. During pregnancies in my late thirties, I experienced this expectation on the part of clinicians. Some authors affirm the importance of having this option. See, e.g., Mary Ann Baily, 'Why I Had Amniocentesis', in Parens and Asch, *Prenatal Testing and Disability Rights*, pp. 64–71; Rapp, *Testing Women, Testing the Foetus*, pp. 3–5. However, whether prenatal testing is truly or fully an option depends on the parties involved, the quality of the relationship between the woman and her physician, the adequacy and accuracy of the information provided, and on the availability of social and economic supports for continuing or discontinuing an affected pregnancy. That younger women may also be influenced by the expectation of clinicians that they undergo prenatal screening is clear from a survey of pregnant women in South Wales. Layla N. Al-Jader *et al.*, 'Survey of Attitudes of Pregnant Women towards Down's Syndrome Screening', in *Prenatal Diagnosis*, 20 (January 2000): 23–29. All of the women were less than 35 years of age. About half were not well informed about the tests, and the majority were unaware that they were voluntary. The few (five out of 101) who refused screening tended to be better educated and of higher social class.
54. The anticipated cost of raising a child who is mentally retarded, and lack of social supports for doing so, constitutes a kind of 'passive eugenics', especially for women or couples whose financial situations are already in jeopardy. Bowman uses the term 'passive eugenics' to apply to the denial of appropriate medical care to the poor. James E. Bowman, 'The Road to

Eugenics', in *University of Chicago Law School Roundtable*, 3 (1997), pp. 491, 493. He implies an inevitable connection between active and passive eugenics: '[A] society that countenances passive eugenics provides fertile ground for both clandestine and overt active eugenics'.
55. Parens and Asch, 'The Disability Rights Critique', p. 13.

Part 3
The Promise of the New Genetics

7
Conditio Humana as Viewed by a Geneticist
Walter Doerfler

Introduction

Thoughts about the human condition have been an important domain of scientists, theologians, philosophers, sociologists and, of course, physicians. In medicine one has to deal with the immediate needs of the individual. Sometimes in physician–patient interactions, enough time does not remain to consider the physician's daily activities in the context of a wider framework. Nevertheless, every practising physician is conscientiously aware of the exposure of the human organism to natural forces within and without the individual patient or healthy human for that matter. Could an interdisciplinary approach somewhere find common ground to help improve our understanding of human nature which has to confront and at least partly comprehend the universe? This is the context in which I will attempt to understand the term *conditio humana*. As humans in this miniscule part of the universe, we are all subject to the conditions defined by the physical parameters of this universe and by the genetics of our biological nature. Environmental conditions may change due to a number of manmade factors and even more factors we have no means of controlling. Moreover, the way in which we relate to theological concepts and religious convictions can profoundly influence our lives. Thus our fortunes are challenged by a complex set of constant and variable determinants which can be defined as the *conditio humana*.

What insights could molecular genetics provide to the advancement of the human condition? Science cannot and will not realistically aspire to change the constraints of nature. But in trying to improve our understanding of basic principles in biology, genetics has contributed to basic medical research and practice. Although the physical well-being of humans will not be the only asset in the 'pursuit of happiness', all human endeavours, perceptions and ideas are sorely dependent on biochemical functions within the human organism. Too materialistic a view? Not at all – the reality of biology and its intrinsic mechanisms set the frame for all physical as well as intellectual and emotional activities in this biology. *Ut sit mens sana in corpore sano*. Whether we prefer to recognize or refute this basic fact of life, the reflections of uncounted molecular and cellular interactions – above all in the human brain – are precondition to human thought, feeling and self-evaluation.

There is probably little merit in meticulous attempts to distinguish between the physical–molecular and intellectual–philosophical aspects of human life. Both are intimately intertwined, and the human brain is a highly sensitive indicator of functional biochemistry and avails itself of a wide gamut of responses to the environment – moods, creative intellectual or physical activities or depressions. Apparently, the human brain is mysteriously capable of intercepting information from the pool of universal reality to a however limited extent. Frankly speaking, our most serious problem lies in the inability to explain these interdependencies in scientific or medical terms. For want of a better term, the word 'pool' will be used here to identify the sum total of all recognizable or incomprehensible facts and laws in the universe including the micro- and macrocosmos and, of course, in the biology as we try to decipher it on the planet earth.

In this chapter, I shall report on the present state of molecular genetics and its applications to the resolution of biomedical problems. Obviously, this text has been written to be comprehensible for a lay audience. Nevertheless, the text will contain several technical terms that will be explained. In the years after 1944, when the molecular basis of all genetic systems was recognized to be encrypted in deoxyribonucleic acid, DNA,[1] rapid progress in the science of molecular genetics has fundamentally altered our outlook on biology, medicine and the human condition. These developments have been greeted as fundamental progress by scientists and physicians pondering important issues of their specialities, although we concede that many of the most urgent questions cannot yet be satisfactorily answered. Outside science, there sometimes remains uncertainty and scepticism *vis-à-vis* the new way of looking at biology from a molecular vantage point. We might be able to overcome misconceptions by helping convey the notion that theology and natural sciences are in fact pursuing similar goals and are asking similar fundamental questions: What is the essence of the universe. What is in the 'pool' of all possible knowledge? To what extent has the new genetics enabled us to reflect more realistically on the human condition?

The now available detailed information on the nature of the human genome, its functions and the genetic bases of human diseases and disabilities will hopefully enable medicine to alleviate human suffering, at least to a major extent. During the past 50 years, medicine has been thoroughly transformed and has reached a level of excellence that previously could not have been imagined. Of course, major problems remain to be addressed on the science side of medicine. At the same time, physicians must be aware of and heed the patients' and the public's apprehensions of an overly technical or purely scientific approach to human ailments and improve their expertise on the personal and psychological needs of their patients. From the patient's point of view, there will always be the need for a clear distinction in attitudes between the scientific and bedside requirements of medicine. Many psychological aspects of medicine have profited as yet only to a limited degree from molecular genetics, mainly due to the appalling complexities inherent in neurobiology. Fortunately, excellent research is being devoted to this challenging field.

In science it is impossible to predict future developments, while scientists in general pursue an optimistic, though patiently sceptical, stand towards the possibilities for further progress. It may be revealing and helpful, particularly to a lay public, to emphasize that we vaguely understand the functional significance of only about 5 per cent of the entire human DNA sequence of 3×10^9 genetic letters. What mechanisms enable the human brain to interact with the physical world in which we have been selected to survive? How are we able to construct in our neurons abstract images of realities in the universe and develop models which render the universe more comprehensible to the evolving human mind: concepts in the religious literature, formulas of physics (e.g. $E = mc^2$), the recognition of the function of DNA? The worldwide wealth in the fields of literature and music further attests to the astonishing but unexplained creativity of the human brain. Which role might information encoded in DNA sequences play in these interactive processes? Future concepts and discoveries will be full of excitement and humble the human mind. The limitations in human brain power, bestowed upon us by evolution at its present stage, define the joys and miseries of the human condition. After all, we have not done so poorly confronting the universe with about 3×10^9 nucleotide pairs in the human genome and 10^{11} neurons plus about 10^{12} glia cells in the human brain.

Basic aspects of biomedicine

Important discoveries in molecular genetics. The identification in 1944 of deoxyribonucleic acid (DNA) as the genetic material[2] in – apart from the RNA viruses - all living organisms has marked the beginning of a golden age of biological, biochemical and biomedical research in, by now, thousands of research laboratories. The rapid progress in the science of molecular biology has fundamentally altered our outlook on biology, medicine and the human condition. While the developments in the field were gradual, though at an unprecedented rate, there have been quantum leaps which have moved the whole field forward. Let me mention just a few of these major discoveries.

The elucidation of the structure of proteins[3] and of DNA[4] facilitated detailed research on the function of these most important molecules in biology.

Proof of the existence of messenger RNA,[5] the intermediate carrier of genetic information from DNA in the cell's nucleus to the ribosomes, the sites of protein biosynthesis in the cell cytoplasm, has helped understand how the genetic information is transported from the site of information storage to the cellular organelles of protein assembly. At the same time, a seminal concept was developed on the principles of the regulation of gene activity in bacteria.[6]

When the primary genetic code that has proved to be universal in all living organisms was deciphered[7] between 1961 and 1966, a major question in biology had been answered. It would not be surprising if DNA, particularly its repetitive sequences, harboured additional, as yet unidentified, codes, but this is pure conjecture at present.

Starting in 1971, the discovery,[8] and later the isolation, of a huge number of bacterial restriction enzymes with the ability to cleave DNA at specific sequences of genetic letters, e.g. at GAATTC, opened the gate and constituted the precondition for the step-by-step analysis of the structure and function of all genomes in biology.

One of the logical and practical consequences of this breakthrough was the development between 1972 and 1974 of sophisticated techniques to clone specific fragments of DNA,[9] again from all organisms. This technological accomplishment marked the beginning of the era of gene technology.

In 1977 two different methods were devised independently by two different laboratories to determine the sequence of nucleotides in DNA.[10] This seminal development initiated the analysis of all genomes at the nucleotide level. Starting in the late 1980s then, the technology of DNA sequencing was upscaled to the determination of the nucleotide sequence in entire genomes of the major model organisms in molecular biology (*Escherichia coli*, many viruses, *Drosophila melanogaster, Caenorhabditis elegans, Arabidopsis thaliana, mus musculus*, several species of apes, *Homo sapiens* and other mammals).

By using a 'simple' viral system (adenoviruses), the 'exploded' organization of most eukaryotic genes could be demonstrated and the splicing mechanism was thus discovered.[11]

It had been known since 1948[12] that some of the nucleotides in DNA were modified. Starting in 1980, it became apparent that the fifth nucleotide, 5-methylcytidine, located at specific sites in the genome had functional significance in the organization of chromatin structure and in the long-term shut-off of gene activities.[13]

The introduction of a modified nucleotide into DNA and specific modifications of the histones, proteins in close association with DNA, form the biochemical basis of the so-called epigenetic phenomenon.[14] Waddington defined 'epigenetics' as 'the branch of biology that studies the causal interaction between genes and their products which bring the phenotype into being'. Epigenetic inheritance then is the transmission of information that is not exclusively encoded in the four-letter genetic alphabet. However, one can also view the role of the fifth nucleotide and the modifications of the histones in a different way. The fifth nucleotide after all is part of the genetic repertoire and can be considered a genetic signal in its own right.

These major discoveries, of which only a few could be listed here, have sparked high-quality research on mechanisms and on the details of many reactions which are essential in the function, replication and genetic makeup of all living organisms. Over several decades now, the concepts and techniques of molecular biology have been applied to the investigation of medical problems, in particular to medical genetics and to the development of new concepts in the diagnosis and treatment of diseases. *Molecular* medicine was born. While many of these endeavours have led to promising avenues of research, many of the major problems in medicine remain unsolved and require further investigations. However, it is hoped that the molecular and genetic approaches to the analyses of medical problems will in the long-run produce new solutions to old problems.

Progress in the basic understanding of biological mechanisms has been paralleled by exciting studies in developmental and cell biology, in epigenetics and in the biology of reproduction. Furthermore, molecular medicine has drawn considerable impetus from the refinement of the knock-in and knock-out technology, the generation of transgenic organisms, the potential of stem-cell research and, starting from 1997, the possibility of cloning mammalian organisms.[15] Fields such as somatic gene therapy and therapeutic cloning have opened up horizons for medicine. These new discoveries have sparked high hopes that might some day bear fruit. In many instances, however, these hopes have proved premature and overrated in their potential, perhaps not in principle but in practice. Nevertheless, research in these areas and its continued financial support will be of paramount importance for further progress in many specialties of medicine.

The nucleotide sequence in the genome of one individual is identical in all of the approximately 10^{13} to 10^{14} cells of the human body. The genome consists of about 3×10^9 nucleotides or genetic letters. The genetic language avails itself of five different letters, A, C, G, T, and 5-mC, the latter a chemical modification of C. In reality, these letters are chemical compounds, nucleotides. Their names are adenosine (A), cytidine (C), guanosine (G), thymidine (T), and 5-methylcytidine (5-mC). For the discussion here, it is not necessary to understand the chemistry of these heterocyclic compounds. The 3×10^9 genetic letters are organized into microscopically recognizable structures, at a first level of resolution into the chromosomes that can be visualized by light microscopy. At the next higher level of complexity, DNA molecules can be made visible by electron microscopy.[16] In the human species, each somatic cell carries 23 chromosome pairs, 22 autosomes and two sex chromosomes. In each chromosomal pair, one allele is inherited from the father, the second allele from the mother. In the germ cells, sperm or egg cells, only one chromosome of each pair is represented. During the maturation of the germ cells, the chromosome number has been reduced to a haploid set, such that after the union of sperm and egg cell, the diploid number of chromosomes is re-established.

It is one of the major enigmas in molecular genetics that among the 3×10^9 nucleotides, less than 5 per cent are actually encoding genes, the carriers of genetic information. There is an estimated number of some 23,000 genes encoded in the human genome. The products of these genes can be modified and rearranged in several different ways. Hence the number of functional gene products is considerably larger than the number of genes. The large majority of nucleotide sequences are made up of often highly repetitive nucleotide sequences that are in large part derived from ancient viral genomes. These now endogenous foreign genomes have entered our genomes at an evolutionarily distant age. The function of these repetitive sequences is not understood. Speculations about their meaning abound; solid information is not available. Some, by no means all, of these sequences can be transcribed into RNA. It will be a difficult, but very important, task to learn more about the biological significance of these sequences. It is inconceivable, and certainly very difficult to explain, that 95 per cent of the human genome, which has to be replicated in its entirety before each cell division at high biochemical expense, has been preserved without an important function in genetics.

Towards molecular medicine. The application of the concepts and techniques of molecular genetics has been particularly successful in the field of human genetics. During the past decades, there have been a large number of excellent publications which have helped clarify the molecular basis of many genetic diseases. Since almost all diseases have a genetic basis, and since genetic predispositions can influence the development and course of most ailments, all specialties in medicine have greatly profited from this novel approach to biomedicine. In earlier decades, the importance of genetics in the diagnosis and treatment of disease was certainly recognized by many, but the means available to implement and apply this information to medicine were still limited. Nevertheless, in many metabolic diseases, e.g. phenylketonuria or Tay-Sachs disease, as well as in various deficiencies in the synthesis of functional haemoglobin, in the so-called haemoglobinopathies, physicians and biochemists have long been aware of the genetic defects underlying these diseases. Many of the important discoveries in basic biochemistry, dating back to the first part of the twentieth century, have decisively shaped medicine and have contributed to the understanding of metabolic and other diseases and, in turn, have had a major impact on genetic research.

Genetic diseases can be roughly subdivided into the following major groups:

1. chromosomal abnormalities
2. monogenic diseases
3. complex, multifactorial or polygenic diseases
4. imprinting defects
5. amplifications of trinucleotide repeats
6. diseases caused by mutations in mitochondrial DNA.

Table 7.1 summarizes the frequencies of possible alterations in the genetic information of humans (occurrence among 1,000 individuals):

Table 7.1

	At age 25	*Lifetime*
Chromosomal abnormalities	1.8/1,000	3.8/1,000
Monogenic diseases	3.6/1,000	20/1,000
Complex or polygenic diseases	46/1,000	646/1,000
Somatic-cell mutations	–	240/1,000

The frequency of somatic-cell mutations is only an estimate. Mistakes accumulate during a lifetime. Many malignancies are causally related to these lifetime mutations, probably in several genes. These mutations are consequences of our environmental exposure to cosmic radiation, metabolic intermediates naturally produced in the human organism, food uptake, viral infections, etc.

1. *Chromosomal abnormalities*: These abnormalities, which can affect the 22 autosomes or the two sex chromosomes – XX in the female, XY in the male – in most instances do not alter the nucleotide sequence of genes and are therefore *in sensu strictiori* not genetic diseases, although they often cause very severe malformations. Chromosomal aberrations are among the most frequent errors during the development of the human embryo with in most cases severe consequences. Some chromosomal aberrations, such as balanced translocations with a partly rearranged but completely balanced genome, leave the carrier without symptoms of any kind. The rearrangements can, however, become a major problem during reproduction when severe misarrangements and imbalances can ensue in the offspring's chromosomes. By transmission of a chromosome carrying an additional segment or a deletion of a segment to the next generation, an imbalanced genotype can arise that will cause severe malformations and/or lead to the premature termination of the pregnancy.

2. *Monogenic diseases*: The genetic diseases caused by mutations (i.e. alterations in, loss or addition of nucleotide sequences), are again subdivided in those caused by mutations in one of the 22 autosomes or in the sex chromosomes. Depending on the biochemical consequences of a mutation, inheritance can be recessive or dominant. In autosomal recessive (AR) diseases, defects manifest themselves only when both chromosomes, the paternal and the maternal alleles, carry mutations. In autosomal dominant (AD) diseases, a mutation on one of the alleles suffices to render the carrier a patient. Since the male complement of chromosomes contains only a single X allele, and since this chromosome is home to a large number of important genes, males can be stricken by severely disabling ailments whenever a mutation falls into a gene residing on the male's only X chromosome.

Thousands of monogenic diseases have already been identified and characterized. Depending on the activity states of other genes, even the monogenic diseases can vary in their penetrance in transmission to the next generation, or their expressivity (i.e. severity of disease phenotype). The monogenic diseases can thus be modified in their severity or age of onset by perhaps numerous other genes, often referred to as modifier genes, but they are actually caused by the mutation(s) in one gene.

3. *Multifactorial diseases*: In contrast, most human diseases are either directly caused or affected by mutations in several human genes. There are probably very few diseases, if any, that do not have a genetic basis. In most instances of diseases due to multifactorial genetic defects, the analysis of the mode of inheritance can be complicated. Predictions about the mode of inheritance in these instances are usually imprecise or impossible.

4. *Imprinting defects*: The term 'genetic imprinting' indicates that the transcription of genes from the paternally and the maternally inherited chromosomes is not identical. In an imprinted region of the human genome harbouring several genes (e.g. A, B, C, D, E, F, G, H) that are present on both chromosomes, genes A, D, E, H will be expressed only from the paternally

inherited chromosome, while genes B, C, F, G exclusively from the maternally inherited chromosomal allele. Thus deletions on the maternal or the paternal chromosome lead to completely different disease phenotypes, and the activities of genes on both the paternal and the maternal chromosomes are required to produce the disease-free phenotype. There are only a relatively small number of genes which fall into this category of imprinted genes. Should there be a defect on one chromosome in an imprinted region, a distinct, frequently severe, disease will ensue. Inheritance of these diseases rarely follows simple Mendelian rules and can be quite complex. Many of these diseases reach a degree of severity which makes reproduction and therefore genetic transmission unlikely. The Prader–Labhart–Willi syndrome and the Angelman syndrome with deletions on chromosome 15, paternal and maternal, respectively, fall into this group of genetic diseases.

5. *Amplifications of trinucleotide repeats*: There are a number of genetic diseases, usually involving the central nervous system, which are due to an amplification of a naturally occurring trinucleotide repeat in different regions of the human genome. In the segments of the human genome that carry such trinucleotide repeats, e.g., $(CGG)_n$ or $(CAG)_n$, there is a normal range for the number n not associated with any disease, e.g. n = 5 to 50 (CGG). This number can be amplified for unknown reasons and by an unknown mechanism to >1,000, and then gives rise to the usually severe disease phenotype. The cause for these amplifications is not understood. Due to the extension of a previously limited and functional nucleotide sequence, the extended version can, depending on its nature, lead to the turn-off of an essential gene or to severe mutations in a gene which cause the disease. The Fragile X Syndrome (CGG), the most frequent cause of mental retardation in human males with only one X chromosome, or Huntington's chorea (CAG, chromosome 4), belong to this group of genetic diseases.

6. *Mutations in mitochondrial DNA*: The mitochondria in the cell cytoplasm provide the machinery to produce the energy for life in every cell. The information required to make up the proteins in the mitochondria is in part stored in a small piece of circular DNA, the so-called mitochondrial DNA, that is different from and encodes functions independent of those in chromosomal DNA in the nucleus. Mutations in the mitochondrial DNA are responsible for very serious genetic diseases. Since the mitochondria are inherited exclusively via the female egg cell, inheritance of these diseases is maternal (i.e. exclusively via the mother).

Incidentally, mitochondrial DNA has been used in studies on the origin of human populations. Some researchers have based far-reaching interpretations on their comparative analyses of the maternally inherited mitochondrial DNA. Recently, it has been claimed that all human populations emanated from East Africa and that, starting perhaps 65,000 (in other estimates 150,000 to 180,000 years ago), a relatively small group of humans from that region have populated and cultivated the rest of the world: the original Eve from Africa.

This brief overview might provide a first insight into the main types of genetic diseases. (There are, of course, many subtleties which cannot be addressed in this short introduction to medical genetics.) In evaluating the information presented here, there are a number of caveats that need be mentioned.

1. There are examples in which the mutation that an individual carries can be directly related to the disease a patient might be suffering from. However, in most instances this interrelatedness can be rather complicated and modified by numerous additional factors. The genetic defect and the disease are related to each other but the connecting mechanism remains to be determined in many genetic diseases.

2. The most frequent diseases – metabolic, tumour, cardiovascular, infectious, psychiatric – involve alterations in several different genes and perhaps in non-coding regions of the genome as well.

3. We are still very ignorant about the possibility of whether alterations in the repetitive nucleotide sequences of the human genome could make a contribution – minor or fundamental – to the origin of disease.

4. There is rapidly accumulating information on the importance of epigenetic mechanisms in the causation of diseases. Epigenetic phenomena are governed, as far as we understand these complex mechanisms today, by alterations in the structure of chromatin and by changes in the positioning of the fifth nucleotide, 5-mC. The biological significance of the latter mammalian genomes has been recognized only about 25 years ago and has not yet been completely elucidated. The unique and specific patterns of distribution of 5-mC and the structure of chromatin and histone modifications are interrelated in an as yet incompletely understood mode. Chromatin forms the biochemically complex protein sheath in and around which DNA is wrapped and organized in the chromosomes. It is conceivable that the causation of frequent and really complex diseases can only be unravelled after we comprehend the organization of chromatin and the meaning of epigenetics at a more advanced level.

The entire human genome as a series of DNA chain molecules – one for each chromosome – presents an extensive array of millions of the four genetic letters A, C, G, T. Upon this array is imprinted a highly specific and conserved pattern of 5-mC nucleotides, the fifth nucleotide. These patterns are inheritable but can differ from cell type to cell type and are different in the different regions of the genome.[17] We do not understand how these patterns are preserved, whether and under what conditions they can change (e.g. in diseases such as cancer) and what their precise functions may be. By now, there is a huge literature on DNA methylation and its biological significance.

Moreover, the correct epigenetic organization of the genome appears to be reprogrammed early in embryonal development, and in its course the patterns of DNA methylation are completely erased and then faithfully re-established. It is questionable whether the circumvention of normal sexual reproduction – by cloning – would not lead to a severe disturbance in the correct epigenetic programming and the re-establishment of the authentic methylation patterns. The frequent malformations in cloned organisms may derive from these deficiencies.

Recent developments and their biomedical implications

The wealth of new information emanating from basic research in molecular genetics has been enthusiastically greeted by scientists and physicians alike, and applied to the study of disease and to the development of novel therapeutic concepts. I can only briefly discuss a few of these recently developed ideas and technologies in biomedicine. In the near future, none of them will be ready for application in the treatment of disease. But basic research on promising new approaches is being conducted in many biomedical laboratories both in academia and the corporate world.

1. *Gene therapy* can be considered at the level of somatic cells or of the germ-line. Since there is at present no reasonable chance to deliver genes to their correct position in the genome of an individual, any attempts to correct a genome in the germ-line would most certainly lead to failure and cause disaster. In cell-culture experiments or with mouse embryo cells, techniques have been developed to target the gene to be delivered into a recipient genome rather precisely. However, many cells have to be used to hit the chosen target successfully in a number of the cells. Germ-line therapy, therefore, remains strictly outlawed. For somatic gene therapy, however, there will be basically two major fields of application: genetic and tumour diseases.

Researchers attempt to introduce non-mutated genes into cells, organs or organisms in which one or several genes have been altered and thus cause disease. The introduced gene coupled to a regulatory region, a promoter, can remain free in the cell nucleus or can insert into the host-cell genome. In the latter instance, the foreign gene may become methylated and inactivated. In the intended treatment of genetic disease, research is aimed at the nuclear transfer and controlled establishment of the non-mutated gene which could hopefully correct the failures due to the presence of a deficient gene. The possible *de novo* methylation of the inserted gene may be prohibitive to success in gene therapy.

In somatic gene therapy against tumours, one hopes to introduce genes that either inhibit tumour growth or make the tumour more easily recognizable by the surveillance by humoral and cellular mechanisms of the immune system. This surveillance mechanism is thought constantly to scan the entire body for the appearance of cells with surface-markers that are recognized as foreign by the immunological control systems or that are sensed outdated since the embryonic stage. While there are hopeful signs of success for gene therapy in many laboratories, most direct applications to humans have failed or have shown unacceptable side-effects. These undesirable effects are at present still too serious to contemplate further immediate application of this technique to the treatment of human tumours. Of course, there are terminal cancer patients for whom risky therapies might constitute the only remaining hope. Basic research will have to continue at a brisk rate.

Particularly when discussing the potential treatment of human tumours by gene-therapeutic methods, one is confronted with the fact that excellent schemes have been developed for the treatment of many human tumours. Gene-therapeutic regimens would have to surpass in efficacy the traditional, highly specialized and optimized therapies already available for the treatment of human cancers.

2. *Stem cells* in the human organism have retained the capability to develop into more specialized cells. Particularly for many important organs, bone marrow, liver, nervous system, heart-muscle, kidney, it would be desirable to use such cells to 'grow replacement parts' for damaged cells or organs. Would it not be wonderful if, after a coronary attack on heart-muscle cells and their destruction with disastrous consequences for the human organism, repaired heart-muscle could be grown in culture starting from the same individual's stem cells? Initial experiments in mice with promising results suggesting the successful replacement of damaged heart-muscle cells (cardiomyocytes) could not, upon critical analysis, be confirmed in other laboratories. In the bone marrow such stem cells have been known for decades and have been used in the therapy of human leukaemias with often excellent results. However, for most other organ systems we are still at the very beginning. We have yet to learn how to obtain, grow and direct the specialization of human stem cells. High hopes and their exaggerated description in the media have led to misunderstandings about the availability of these futuristic therapies. Stem-cell research is still in its infancy, but is vigorously pursued in many laboratories.

The possibility of using more easily attainable and functionally supposedly more pliable human embryonic stem cells has evoked a worldwide debate about complex ethical implications. The lack of acceptance in the general public *vis-à-vis* the generation and use of embryonic stem cells for the potential treatment of diseases has rendered it difficult (and in some societies impossible) to carry this possibility further. Hopefully, one will learn more about the application of adult stem cells from a patient's own body for use in the treatment of human disease. Of course, even in societies that presently refute the use of human stem cells for research, attitudes may change with time and with the realization that new medical technologies will help to alleviate suffering in the ageing societies of the developed countries. For the physician and medical geneticist, this latter aspiration must remain one of the top priorities, along with important ethical considerations. Hence our viewpoints might differ greatly depending on one's primary professional affiliation. The physician's highest priority, however, must always remain *primum nil nocere*.

Human stem-cell research might realistically have to concentrate on the development of adult stem cells for organ repair: at present a remote goal. There are open questions galore, and several of the highly acclaimed results from some laboratories have not been reproduced in others. Basic research on simple model systems seems to remain the order of the day.

3. In 1997 the generation of a cloned sheep in Scotland[18] (the 'Dolly' incident) has also made possible an improved understanding of human development. Previously it was thought that the non-sexual reproduction of mammalian organisms was impossible. Dolly, and since then many different mammalian organisms, were presumably generated by transferring the nucleus of a highly differentiated cell of the adult body into an egg cell from which its own nucleus had been removed. While such procedures were already successfully carried

out in amphibians more than 30 years ago, Dolly opened new possibilities for mammalian organisms in spite of the poorly understood, difficult to control, epigenetic complications. At this time, it appears likely that only a small subset of the body's cells persists in a developmental state compatible with the functionality of their nuclei in these nuclear-transfer experiments. Moreover, among cloned animals a high proportion have been afflicted by severe malformations and a limited life-span.

4. For practical purposes, one distinguishes between reproductive and therapeutic cloning. There is worldwide agreement that reproductive cloning is and will be absolutely unacceptable for humans. Apart from an ethical point of view, the peril of malformations rules out further thought about such procedure. A totally different problem arises from plans to apply the new, still incompletely understood, technology to generate early human embryos or to use early embryos from *in vitro* fertilization procedures or perhaps aborted embryos for the development of human organs. While essential elements of the biology of cloning of organisms will still have to be worked out, important ethical controversies will have to be considered. Their resolution in society appears a precondition to proceed further. In the meantime, the basics could be elucidated by using mouse models, although it seems clear that this model system can help resolve some, but probably not all, of the important remaining questions for human biology and medicine.

The new biology and society

In all fields of biology and particularly in medicine, the breakthroughs in biological research have been welcomed as a tool to study complex problems of immediate relevance for clinical medicine. This enthusiasm should not entice us to ignore the magnitude of intellectual and practical problems remaining before results in basic molecular biology can be applied to the treatment of disease and disability. The gateway to increased knowledge and progress in clinical medicine will continue to be top-quality basic research.

Confronted with the results of research in molecular biology, the public has, at least initially, reacted with scepticism and even apprehension. These attitudes were sometimes enhanced, even exploited, by politicians and some of the media. Unrealistic risk assessments emanating from scary scenarios of exaggerated dimensions have blurred many people's vision. Since I have been working predominantly in Germany for the past 34 years, my perceptions in this respect may have been honed by an overcritical public which had to cope psychologically with the country's traumatizing history in the first part of the twentieth century. The country's past could not be accommodated without highly sensitive reactions in the heir generation, in particular towards possibilities of genetic manipulation of the human organism. For scientists exposed to this societal environment, careful and sensible information *vis-à-vis* the public, particularly in the younger generation, has to be the order of the day. My own experiences, after more than 150 lectures to the general public, have left me optimistic. Medically relevant information has always received interested acceptance. However, under the influence of some of the more opportunistic politicians, legislation has been passed in Germany which is damaging particularly green-gene technology. We can only caution our colleagues in more sedate societies to be aware of the dismal consequences of legislation aimed at restricting scientific research. Once a law has been passed, it is very difficult to change or to abolish it. Moreover, a field of research stigmatized by restrictive legislation will invariably arouse suspicion even in the well-meaning part of society. My generation is particularly alerted, because Germany lost some of her best scientific talent in the 1930s and 1940s. The political class, perhaps not only in Germany, still appears incapable of recognizing and setting priorities commensurate with the magnitude of problems in higher education and scientific research.

Questions of eugenics, though actively entertained in several countries during the early decades of the twentieth century, were carried to a criminal extreme in Germany between 1933 and 1945. In my generation, therefore, we can approach this topic only with profound caution and intellectual caveats. Aside from possible ethical and religious objections, there is not a shred of evidence for the existence of tools which could facilitate a directed and regulated improvement of any genome, let alone the human one. Scientists are fully aware of the immense complexities of biology and can only advise against any attempts or speculations on eugenic measures.

The second most important task for scientists, next to their duty to search for novel insights into nature and the recognizable reality of this universe, amounts to informing the public in comprehensible terms about their discoveries. Science in general and biomedicine in particular, with its immediate connotations for human health and well-being as well as for the acceptance of disease and malformation in society, will suffer and lose long-term sustainable support if scientists neglect to take the general public along on their never-ending journey into the unknown. It is the public that supports the majority of scientific research, and even private funding is connected in many ways to public accomplishments. It will be difficult to attain a general consensus on many of these complex issues without improving the general public's standard of information on modern biology. Each society, with its specific historic sensibilities, and each culture or religion, will have its own ethical and cultural aspirations and convictions. Each individual, even in a somewhat homogeneous society, will develop her or his own conclusions when it comes to the judgement of complex issues related to medical research.

Conclusion

The joy of science is in the journey, not in the arrival. This conundrum might serve to summarize not only the joy but the reality of science. Most questions asked in any scientific endeavour can only hope to receive a transiently valid answer. Tomorrow, new techniques, new ideas, could render today's accomplishments obsolete. However, without today's concepts, future developments will not evolve and mature.

To what extent has the new genetics enabled us to shape a more realistic image of the human condition? Many generations have asked, and will continue to ask, these and other more important questions. As we proceed with research, we will gain new insights and yet understand that the answers to the most difficult questions will remain outside the realms of our comprehension. Will we forever be limited by our present repertoire of 3×10^9 nucleotides and 10^{11} neurons?

Nevertheless, it remains amazing that the brain with 10^{11} neurons is capable of pondering the nature of life and the universe. By what mechanism can the human brain at all partake at least in part in the pool of all possible information? There must be a biochemical correlate for this share-enabling interaction that might somehow be dependent on multiple interactions between interconnected neurons. But what functions inside the neurons? Could the billions of nucleotides, their sequence, their surface-charge patterns along the chain molecules, serve as the receiving antenna for this interaction with the pool of all possible information? Particularly the repetitive DNA sequences might carry additional as yet unrecognized coding principles. New ideas are required to elucidate these unknown mechanisms. Speculation can only be the beginning of serious research. A century ago, scientists were wondering about the storage of genetic information, about the genetic code, and had no idea of the pliable functions of nucleic acids or proteins. Should these classes of molecules be active in yet additional ways obscured in today's science as a conundrum of gigantic proportions?

Notes

1. O.T. Avery, C.M. MacLeod, M. McCarty, 'Studies on the Chemical Nature of the Substance Inducing Transformation of Pneumococcus Types. Induction of Transformation by a Deoxyribonucleic Acid Fraction Isolated from Pneumococcus Type III', in *Journal of Experimental Medicine (J Exp Med)*, 79 (1944), pp. 137–58.
2. Avery *et al.*, 'Studies on the Chemical Nature'.
3. L. Pauling and R.B. Corey, 'Stable Configurations of Polypeptide Chains', in *Procedures of the Royal Society of London, Series B, Biological Sciences (Proc R Soc Lond B Biol Sci)*, 141 (1953), pp. 21–33.
4. J.D. Watson and F.H. Crick, 'Molecular Structure of Nucleic Acids; a Structure for Deoxyribose Nucleic Acid', in *Nature* 171 (1953), pp. 737–38.
5. F. Gros, H. Hiatt, W. Gilbert, C.G. Kurland, R.W. Risebrough and J.D. Watson, 'Unstable Ribonucleic Acid Revealed by Pulse Labelling of Escherichia Coli', in *Nature*, 190 (1961), pp. 581–85.
6. F. Jacob and J. Monod, 'Genetic Regulatory Mechanisms in the Synthesis of Proteins', in *Journal of Molecular Biology*, 3 (1961), pp. 318–56.

7. M.W. Nirenberg and J.H. Matthaei, 'The Dependence of Cell-free Protein Synthesis in E. Coli upon Naturally Occurring or Synthetic Polyribonucleotides', in *Procedures of the National Academy of Sciences, USA (Proc Natl Acad Sci USA)*, 47 (1961), pp. 1588–602.
8. T.J. Kelly and H.O. Smith, 'A Restriction Enzyme from Haemophilus Influenza II', in *Journal of Molecular Biology*, 51 (1970), pp. 393–409.
9. D.A. Jackson, R.H. Symons and P. Berg, 'Biochemical Method for Inserting New Genetic Information into DNA of Simian Virus 40: Circular SV40 DNA Molecules Containing Lambda Phage Genes and the Galactose Operon of Escherichia Coli', in *Proc Natl Acad Sci USA*, 69 (1972), pp. 2904–9; P.E. Lobban and A.D. Kaiser, *Journal of Molecular Biology*, 78 (1973), pp. 453–71; J.F. Morrow, S.N. Cohen, A.C. Chang, H.W. Boyer, H.M. Goodman and R.B. Helling, 'Replication and Transcription of Eukaryotic DNA in Escherichia Coli', in *Proc Natl Acad Sci USA*, 71 (1974), pp. 1743–7.
10. A.M. Maxam and W. Gilbert, 'A New Method for Sequencing DNA', in *Proc Natl Acad Sci USA*, 74 (1977), pp. 560–64; F.F. Sanger, S. Nicklen and A.R. Coulson, 'DNA Sequencing with Chain-terminating Inhibitors', in *Proc Natl Acad Sci USA*, 74 (1977), pp. 5463–7.
11. L.T. Chow, R.E. Gelinas, T.R. Broker and R.J. Roberts, 'An Amazing Sequence Arrangement at the 5' Ends of Adenovirus 2 Messenger RNA', in *Cell*, 12 (1977), pp. 1–8; S.M. Berget, C. Moore and P.A. Sharp, 'Spliced Segments at the 5' Terminus of Adenovirus 2 Late mRNA', in *Proc Natl Acad Sci USA*, 74 (1977), pp. 3171–75.
12. R.D. Hotchkiss, 'The Quantitative Separation of Purines, Pyrimidines, and Nucleosides by Paper Chromatography', *Journal of Biological Chemistry (J Biol Chem)*, 175 (1948), pp. 315–32.
13. D. Sutter and W. Doerfler, 'Methylation of Integrated Adenovirus Type 12 DNA Sequences in Transformed Cells is Inversely Correlated with Viral Gene Expression', in *Proc Natl Acad Sci USA*, 77 (1980), pp. 253–56; A. Razin and A.D. Riggs, 'DNA methylation and gene function', in *Science*, 210 (1980): 604–10; W. Doerfler, 'DNA Methylation and Gene Activity', in *Annual Review of Biochemistry (Annu Rev Biochem)*, 52 (1983), pp. 93–124.
14. C.H. Waddington, 'The Epigenotype', in *Endeavour*, 1 (1942), pp. 18–20.
15. I. Wilmut, A.E. Schnieke, J. McWhir, A.J. Kind and K.H. Campbell, 'Viable Offspring Derived from Foetal and Adult Mammalian Cells', in *Nature*, 385 (1997), pp. 810–13.
16. A.K. Kleinschmidt, D. Lang, D. Jacherts and R.K. Zahn (1962), 'Preparation and Length Measurements of the Total Deoxyribonucleic Acid Content of T2 Bacteriophages', reprinted *Biochemica et Biophysica Acta*, 1000 (1989), pp. 41–48.
17. W. Doerfler, *Foreign DNA in Mammalian Systems* (New York: Wiley-VCH, 2000).
18. Wilmut *et al.*, 'Viable Offspring'.

8
Researching Genetics and Health: Implications for Public Health Primary Care Medicine
Blair Smith

Introduction

To be human is to experience illness – that is normal. Health and illness form a spectrum, rather than a dichotomy, as do normality and abnormality.[1] Few, if any, of us are completely free of 'disability' (for example I wear spectacles) or 'abnormality' (for example I am left-handed). The point on the health–illness spectrum at which an individual or professional determines that intervention may be appropriate varies according to many factors, including previous experience, culture, beliefs and the (perceived) availability of effective treatment. An equally normal part of being a human is the desire to provide treatment and health care for people identified as having an illness or disability. Members of the medical and related professions are those with the vocation, qualifications and remuneration to provide formal health care, but the duty probably belongs to all members of society. From the beginning of the historical chapter in which humans made their first appearance, the latest technology and information have been pressed into this service. This is part of what allows human society to advance, and the conduct and application of medical research are therefore among our duties on behalf of our future generations. We are also duty-bound to ensure, as far as possible, that the conduct and impact of this research are fully considered, ethical and justifiable (and this should begin by a full consideration of the meanings of these concepts). Genetic science is no different from any other technology in these respects, and seeks above all to inform the design and provision of medical treatments.

Since the recognition of the importance of Mendel's work,[2] research into the genetics of health and illness has produced important and interesting results that have informed us about the human species and about certain of its illnesses. This provides new insights into what it means to be a human, in a biological sense, and into how the illnesses that help to define our humanity develop. Genetic science has always been accompanied by controversy and has provoked detailed debates around the associated ethical implications. It is important that medical scientists are familiar with these debates and their purposes (which are, among other things, to remind us of our obligations and to assure our humanity). It is equally important that ethicists are familiar with the aims of genetic science, which are the further understanding of humans, and the treatment of their diseases. The elimination of disability is neither an aim nor a possibility (without eliminating the human species), but the alleviation of suffering, through genetic science, is both.

DNA and genetics

Our genes form the 'blueprint' for who or what we are, biologically. Genes are sections of the chromosomes that exist in the nucleus, or centre, of every cell in our body. In all but a very few cases we have 46 chromosomes per cell, each made up of tightly wound DNA (deoxyribonucleic acid), in the famous double helix structure first reported by Watson and Crick in 1953.[3] We inherit two copies of each chromosome, one from each parent, and therefore some characteristics from each. The precise sequence of nucleic acids in the sections of DNA that form each gene determines their capacity to code for, and allow the manufacture of proteins. It is these proteins that initiate and are involved with metabolic processes, form the constituents of every cell in the body and are the 'building-blocks of life'. There are only four different nucleic acids in DNA – adenosine (A), cytosine (C), guanine (G) and thymine (T) – but sufficient variations in the order in which these occur in the DNA molecule to build all of the proteins we need. A slight variation in the order of nucleic acids in the gene – perhaps a single change, such as a C instead of G – may alter the resultant protein such that there is a noticeable difference in its function or effect. In some cases, this may cause or contribute to a disease or a disability. These differences, or variants, may be the result of a pattern inherited from one or both parents, or the result of a spontaneous change, or mutation in the individual's gene. Given that there are some 3,000 million nucleic acids in each cell, the search for these variants is challenging, to say the least. Other genetic causes of disease or disability may be more obvious, such as an extra or absent chromosome, or section of DNA in a gene.

The twentieth century saw major advances in our understanding of human biology at this molecular level. Now in the twenty-first century two important new factors offer society important opportunities for major advances in genetic science at a population level, and the need for parallel philosophical debate is at least as great as ever.

1. *The mapping of the human genome.* In June 2000 the International Human Genome Sequencing Consortium and Celera Genomics Corporation announced the completion of the Human Genome Project.[4] This publicly funded, multinational project had aimed to unravel and identify the sequence of the nucleic acids that occurs in the DNA of every human chromosome, allowing the identification of genes. The results are available as a resource for scientists on several websites, such as that of the European Bioinformatics Institute. As a result of this, the full complement of human protein-coding genes (some 30,000) are either identified already, or are likely to be soon.[5]

2. *Developing scientific methods.* (a) There is now widespread availability of cheap, rapid laboratory techniques for detailed analysis of DNA. This allows the DNA of large numbers of blood samples to be analysed in great detail, examining gene sequences and identifying variants efficiently and precisely. (b) The measurement of genetic contribution to disease and disability is extremely complex. Laboratory developments therefore need to be matched by methods of analysing and interpreting data from complex systems, and this new science of *genetic epidemiology* is developing rapidly.[6]

Genetic research and clinical practice

Until recently, genetic research focused on important but relatively uncommon diseases with classic Mendelian inheritance patterns. These include, for example, cystic fibrosis, phenylketonuria and sickle cell disease. These conditions cause considerable disability in individuals, and are important in societies where the prevalence of the specific genetic variant is high. This has therefore been important research. Study of these conditions has also contributed greatly to our understanding of genes and of modes of genetic inheritance, and has facilitated further research into the causes of inherited disease.

However, in most societies, and in most areas of clinical medical practice most of the diseases with well-described genetic patterns are rare, and their overall impact relatively low. Genetic research has therefore had little impact on the public or on primary care general medical practitioners (known as GPs, or family practitioners), and there has been little need for either group to understand its outcomes or implications. More recently, a deeper understanding of the heritable component of some diseases that have multifactorial causes has been achieved. For example, some types of breast cancer have been found to have an important genetic contribution, associated with the BRCA1 and BRCA2 genes.[7] A woman's family history of breast cancer alone is a poor predictor of her own outcome (i.e. whether or not she will develop breast cancer). Family history together with analysis of these genes allows much more accurate prediction, and raises the possibility of preventive treatment. Known genetic causes of breast cancer remain, however, relatively uncommon, and the great majority of cases arise without these genetic mutations being present. Similar 'Mendelian subsets' with highly penetrable genetic variants have been discovered for other common conditions, including diabetes, osteoporosis and bowel and ovarian cancer. This is an important area of continuing research which will help us to identify individuals at risk of developing these serious diseases at a stage where preventive treatment may be possible.

However, we can now increase our scope even from these developments, as a result of the human genome map and the emerging laboratory and statistical techniques. Most medical conditions and causes of disability have multiple causes, many of which remain poorly understood. At its broadest, these can be divided into 'genetic' and 'environmental' causes. The former relate to abnormalities (to use the negative term that informs the theme of this book) or variations in DNA or chromosomes; the latter relate to everything else, including smoking, lifestyle, life events and atmospheric pollution. These have also been termed, respectively, 'nature' and 'nurture'. For example, the risk of ischemic heart disease (IHD – the cause of angina and myocardial infarction) has a clear inherited component but is also strongly associated with smoking, a diet high in saturated fat and lack of exercise, among many other environmental factors. The genetic factors, although important, remain largely unknown. This is probably because

1. there is unlikely to be a single genetic factor that causes IHD;
2. such genetic factors as may be identified are each likely to contribute only a small proportion of the cause of IHD; and
3. the interaction between any genetic factors and the environment in causing IHD is likely to be very complex.

However, an understanding of the genetics of IHD and the interaction with environmental causes would be very useful both in preventing the disease in susceptible individuals (by targeting preventive treatment appropriately) and understanding the relative importance of modifiable environmental causes (leading to improved public and individual risk of disease). Much effort and resource is currently invested in UK primary care in detecting and treating high levels of cholesterol in the blood, an important risk factor for IHD. If this effort could be restricted to those whose genetic constitution made their overall risk of IHD significant, this effort and resource could be redirected to other areas of need, and many unpleasant side effects of cholesterol-lowering treatment could be avoided.

Genetic epidemiology

Epidemiology is 'the study of the distribution and determinants of health-related states or events and the application of this study to control health problems',[8] and genetic epidemiology is 'the study of the role of genetic factors and their interaction with environmental factors in the occurrence of disease in human populations'.[9] Genetic epidemiology studies aim to identify the genes associated with specific diseases and disabilities, and the relative contribution of variants of these genes to the development of the conditions in individuals and in society. Broadly, there are two kinds of genetic epidemiology study: (1) gene-discovery studies, which aim to identify specific genes and their variants; and (2) gene characterization studies, which aim to measure the importance and behaviour of these inherited causes of disease and disability (see Box 8.1, pp. 139–40).

Early genetic epidemiology studies were based on small, specific groups of people, such as those with a specific disease, or family or community groups. These allowed the efficient identification of relatively straightforward aetiologies, where the genetic contribution was strong, or the disease outcome easily measured. However, for most diseases, such as IHD, the aetiology is more complex and occurrence harder to identify reliably. Furthermore, it is important to be able to apply the findings of this research to the general population, and findings from isolated or specialized populations cannot be extrapolated beyond these groups.

Increasingly, therefore, the need for much larger study samples is recognized.[10] This recognition is in parallel with a realization of the potential importance of this research, and consequent national and international funding. Across the world, large and very large population-based genetic studies are being established, with two broad aims: to identify genes and their variants that are associated with a wide range of illnesses (including common and complex diseases) and responses to treatments; and to assess the relative genetic and environmental contributions to the development of these illnesses and responses (see Table 8.1, pp. 139–40). In this country, initiatives such as the UK Biobank,[11] and Generation Scotland[12] are among the most prominent initiatives internationally. The former is a study funded by the Medical Research Council, the Wellcome Trust, the Department of Health and the Scottish Executive Health Department (SEHD), and aims to recruit 500,000 adults aged between 40 and 69 with a view to studying gene-environment interaction in the aetiology of common complex diseases. The latter initiative includes the Scottish Family Health Study, which is funded by SEHD and aims to recruit up to 50,000 adults, including as many members of each participating family as possible, with a view to studying the heritable factors associated with diseases, disease-risk factors, and response/non-response to drug treatments. Both of these studies are currently in their early phases. They will measure a number of lifestyle and clinical factors, assess family medical history, and take blood samples for DNA analysis from all participants.

Research outcomes

This research will have the scientific gain of improving the biological understanding of our species and its health and illness. The ultimate clinical prize is the development of medical treatment and prevention strategies based on the new genetic knowledge and targeted appropriately – 'personalized medicine'. For example, a genetic test or a battery of tests might determine an individual's risk of specific diseases and their likelihood of responding to specific drug therapies. These already exist for some conditions. There are, for example, several prenatal genetic tests that are performed frequently, even routinely, including those for Down's syndrome and cystic fibrosis. For adults, tests are also available for Huntington's chorea and for the genes associated with breast cancer.

In practice, however, the goal of truly personalized medicine may be somewhat distant for a number of reasons,[13] and a healthy degree of scepticism is appropriate. Nonetheless, early promised outcomes of this research will still be valuable, and include:[14]

1. Improved understanding of the aetiological gene-environment interactions for many of the major conditions (for example, heart disease, diabetes and cancers). This will allow physicians and their patients to focus on the modifiable, environmental risk factors with an understanding of the effect this will have in reducing risks.
2. Increased genetic testing, from pre-conception onwards. This will allow those who want either to minimize the risk of certain diseases or causes of disability, or to prepare for their impact in advance.
3. Describing new taxonomies of pathophysiology and disease, based on molecular classifications rather than signs and symptoms, with distinct information about prognosis and treatment.[15] This should lead to a more precise understanding of the causes of disease and disability, leading to more precise treatments and preventive strategies.
4. Targeted prevention and prognosis ('dia-prognosis')[16] based on genetic or molecular factors. This will allow better information to be made available to individuals who have, or who are at risk of, disease and disability.
5. Pharmacogenetics – the science that studies genetic differences that explain individual variations in response to drug therapies, or the development of adverse effects. This will lead towards individually targeted therapeutic drug strategies.
6. Determining the heritability and familial aggregation of illnesses or disease risk factors, and thus directing further gene discovery studies and maximizing the potential of and for the above outcomes.

Some important considerations

This optimism must, however, be tempered both by realism and attention to the ethical, legal, social and spiritual issues that the science must generate. These are the main subject of this book and are not the focus of this chapter. An important early problem, though, will be the time-lag between the discovery of genetic causes of disease and the ability to treat or modify these causes. This interval is unpredictable, but may be measured in decades or longer for many important diseases or causes of disability. An important result will therefore be that quantified genetic risk may be conferred upon individuals without the ability to reduce that risk. In prenatal medicine, the option of termination of pregnancy is often chosen (though should not be pushed by the medical profession) when disease or disability is diagnosed or discovered to be a high risk. This rather drastic option is not, of course, available after birth. The anxiety generated by a positive result is already well-described.[17] For example, for an individual with a family history of Huntington's chorea – a very disabling condition with onset in late middle age and for which there is no cure – distress is the most likely reaction to the discovery that he or she carries the relevant gene and will inevitably develop the condition. This time-lag should not, in itself, prevent the research being undertaken, for the findings are required before development of the cures can become possible. It means, though, that scientists, professionals and the public must have a clear understanding of the potential problems that will be raised, and develop strategies to manage them. It also means that those conducting and participating in the research must do so for largely altruistic reasons, rather than for immediate evidence of its benefits.

For some conditions, treatment is currently available on the discovery of genetic risk. Discovering the BRCA gene mutations and their importance in predicting breast cancer has led to the offer of prophylactic mastectomy and/or oophorectomy for women with a relevant family history and one of these mutations. This procedure leads to a gain in years of life, albeit modest,[18] though it involves radical and disfiguring surgery. Development and introduction of gene therapy to treat conditions such as this remains a distant vision.

Implications for primary medical care

Another issue is the widespread lack of knowledge and understanding of genetics by the public and the health professions (including, in particular, primary care), and a lack of educational and financial resources to deal with this.[19] Although at present the workload created for most GPs by genetic issues is low,[20] the potential workload is high, now and in the future. For example, an average GP in the UK, with a list size of 1,700, has been estimated to have 140 registered patients with a strong family history of breast, ovarian or colorectal cancer, of whom 56 express concern about this when questioned, but of whom only ten could be referred to clinical genetics based on current guidelines.[21] The time and expertise required to advise and support those who cannot be referred to specialist services is considerable but does not generally exist. As the output of genetic science increases, as more genetic tests become available, and as more individuals become affected by their results, we will need to develop the financial, human and educational resources to deal with this growing shortfall in meeting expressed health needs. Research on risk within families suggests that there will be considerable increased workload in primary care created by individuals anxious about high perceived risk of disease or disability but low actual risk – the 'worried well'.[22] Again, this will increase as more information on family risk becomes available with the identification of genes for specific diseases and risk factors.

Barriers identified to the provision of genetic services in primary care include inadequate knowledge of basic genetics, lack of detailed or current family histories, lack of confidence and lack of referral guidelines,[23] ethical dilemmas associated with the therapeutic gap in the context of common cancers[24] and relative perceived unimportance beside other practical and demand-led priorities.[25]

Specific philosophical and spiritual issues surrounding predeterminism, prenatal screening and the potential for selection, and the uncovering of previously unknown familial relationships, will need to be addressed by society, and the medical profession must be part of that society. At the frontline of the medical profession, GPs will very soon need to be equipped with the information, strategies and language to communicate with their patients and professional colleagues about these issues. This will be a new side to the provision of primary medical care for many practitioners.

Conclusion

Genetic science offers the potential to make important advances in the design and delivery of medical care. With care at this stage, today's research can be a major resource for tomorrow's generations. There are several pitfalls we must avoid, however, and their identification must be a priority, leading to informed research and debate. These must be pursued with the same vigour as the laboratory and clinical science. Every new technology has the potential to cause benefit and harm. The medical and scientific professions are as keen as any to ensure that the risk of harm is minimized, while the chances of benefit are maximized. This will require powerful collaborations between doctors, scientists, philosophers, ethicists, social scientists, theologians, artists and, of course, lay people – in short, all of humanity! In this way, genetic science can help us to understand and celebrate the diversity of humanity, in the way that I believe most of my professional colleagues would like to.

Box 8.1 Genetic epidemiology study types.

1. Gene discovery studies	
Association studies	Testing the association between known or suspected 'candidate' gene alleles or variations (polymorphisms) and specified outcomes (phenotypes). Polymorphisms may be as precise as a change in a single nucleotide of the relevant DNA sequence – a single nucleotide polymorphism (SNP).
Linkage studies	Searching for gene loci (positions) that are linked with specified outcomes. Variations at gene loci are identified by their higher than expected frequency among relatives who share phenotypes, such as a disease. They may be specific alleles that are directly related to the molecular cause of the disease, or 'gene markers' – other alleles that do not cause the disease, but that are inherited more frequently with the causative allele because they are in 'linkage disequilibrium'. Linkage studies need to be based on family units. They are less precise than association studies, but may allow examination of multiple loci/markers at once, for further testing by fine mapping studies or association studies.

2. Gene characterization studies. These research:	
Heritability or familial risk of diseases or traits	Family studies that recruit subjects through a 'proband' with a specific disease employ case-control or cohort methods, aiming to calculate the recurrence risk of that disease within the family. The risk will depend on the type of relative and the heritability of the disease, but the actual gene(s) involved do not need to be known.
Segregation: identifying the mode of inheritance of familial diseases or traits	Examining data on disease occurrence within families will identify the probable mode of inheritance. This is relatively simple for the (rare) conditions with simple Mendelian inheritance caused by highly penetrable genes, but more complex for other traits, such as cardiovascular disease. Most studies will lack the power to identify the mode of inheritance for these complex traits
Gene – environment interactions	Quantifying the relative gene: environment influence on disease or trait aetiology. Distinguishing between these two major factors requires complex statistical procedures and detailed data on genotypes and phenotypes of large numbers of people, either in cross-sectional or cohort studies. Genetic and environmental factors may act independently or interact with each other. Twin studies are a powerful means of examining gene – environment interactions. At its simplest, in a group of twins where each pair was raised together, any differences in the rate of phenotypic expression between monozygous (identical) and dizygous twins can be assumed largely to represent environmental aetiology.

Table 8.1 Large-scale genetic epidemiology studies planned or underway[26]

Study location	Study name	Sample size (proposed)	Sample characteristics	Stated objectives
Australia	Western Australian Genome Health Project	2,000,000	All consenting members of Western Australian population (total 2,000,000), linked with existing cohort studies, routine health records and 'biospecimen banks'.	To investigate the changing roles of genes, environment, gene – gene, and gene – environment interactions over the entire life-span.
Canada	CARTaGENE	60,000	Representative of the population of Quebec, aged 25–74.	To study the founder effect and ethnic heterogeneity.
China	Kadoorie Project[27]	500,000	Representative sample across several regions of China.	To study gene-environment interaction in the aetiology of common diseases, and to serve as a resource for future research.
Estonia	Estonian Genome Project	100,000	Volunteers, during routine primary medical care (no stated age restriction).	To create a database of health, genealogy and genome data that will comprise a large part of the Estonian population, aiming for about 71 per cent of the population.
Iceland	DeCode/ Icelandic Health Sector Database	275,000	The population of Iceland, unless individuals opt out. Routine data and samples.	To map the genes of the Icelandic population, and to investigate twelve genetically influenced diseases.
Latvia	Genome Database of the Latvian population	None stated	Volunteers (no age restriction), during routine primary medical care, aiming for a nationally representative sample.	To build a genome database of the Latvian population, with multiple clinical and commercial expectations.
Mexico	INMEGEN	None stated	Individuals in six remote regions of Mexico.	A 'race-based genome project', the goal of which is to glean insights into genetic differences believed to be unique to its population that may

Table 8.1 (continued)

Study location	Study name	Sample size (proposed)	Sample characteristics	Stated objectives
				play a key role in chronic diseases like asthma, diabetes and hypertension.
Sweden	UmanGenomics	100,000	Inhabitants of Västerbotten, northern Sweden.	Database of genetic and medical information, for commercialisation and academic research.
Sweden	LifeGene	500,000	Maximum age 55, enrolled through national databases and publicity campaigns.	To improve public health through providing an open resource to researchers. There will be a focus on disease aetiology and everyday health problems, including important chronic diseases and the possible role of infections as risk factors.
Singapore	Genome Institute of Singapore	None stated	The population of Singapore.	To establish a database of information on personal medical history, genealogical and genetic data.
UK	UK Biobank	500,000	Adults aged 45–69. Recruitment method to be finalized, but probably involving primary care.	To study gene-environment interaction in the aetiology of common diseases, and to serve as a resource for future research.
UK (Scotland)	Generation Scotland: the Scottish Family Health Study	50,000	Groupings of families (first-degree relatives aged 18 or over) including at least one sibling pair in each family.	Assessment of the familial aggregation and heritability of important disease-related traits, and identification of genetic loci that contribute to these traits.
USA[33]	Not yet determined	500,000	Not yet determined.	To study genetic and environmental influences on common

Table 8.1 (continued)

Study location	Study name	Sample size (proposed)	Sample characteristics	Stated objectives
				diseases. Currently a preliminary proposal only.
Multi-national	GenomEUtwin	850,000	Twin pairs: Finland, Sweden, Norway, Denmark, The Netherlands, Italy, UK, Australia.	To characterize genetic, environmental and lifestyle components in the background of health problems.

Source: Table adapted from Smith *et al*. 'Genetic Epidemiology and Primary Care', with permission. Details correct at time of writing. For further information, see http://www.p3gobservatory.org/welcome.do

In addition several national and multinational projects are underway to combine the resources of a number of existing cohorts and specimen banks for genetic epidemiological research.

Notes

1. R.J. Taylor, B.H. Smith and E. Van Teijlingen (eds), *Health and Illness in the Community – an Oxford Core Text* (Oxford: Oxford University Press, 2003).
2. G. Mendel, 'Versuche über Pflanzen-hybriden', in *Verhandlung des naturforschenden Vereines in Brünn*, 4 (1866), pp. 3–47.
3. J.D. Watson and F.H.C. Crick, 'Molecular Structure of Nucleic Acids', in *Nature*, 171 (1953), pp. 737–38.
4. International Human Genome Sequencing Consortium: 'Initial Sequencing and Analysis of the Human Genome', *Nature*, 409 (2001), pp. 860–921; J.C. Venter *et al.*, 'The Sequence of the Human Genome', *Science*, 291 (2001), pp. 1304–51; International Human Genome Sequencing Consortium: 'Finishing the Euchromatic Sequence of the Human Genome', *Nature*, 431 (2001), pp. 931–45.
5. European Bioinformatics Institute: www.ensemble.org; International Human Genome Sequencing Consortium: 'Finishing the Euchromatic Sequence of the Human Genome', pp. 931–45.
6. M.J. Khoury, J. Little and W. Burke (eds), *Human Genome Epidemiology* (New York: Oxford University Press, 2004).
7. Y. Miki, J. Swensen, D. Shattuck-Eidens, P.A. Fureal, K. Harshman and S. Tavtigian, *et al.*, 'A Strong Candidate for the Breast and Ovarian Cancer Susceptibility Gene BRACT', *Science*, 266 (1994), pp. 66–71; R. Wooster, G. Bignell, J. Lancester, S. Swift, S. Seal and J. Mangion, *et al.*,

'Identification of the Breast Cancer Susceptibility Gene BRCA2', *Nature*, 378 (1995), pp. 789–92.
8. R.J. Last, *A Dictionary of Epidemiology* (Oxford: International Epidemiological Association, 4th edn, 2001).
9. Last, *A Dictionary of Epidemiology*.
10. A. Wright, A. Carothers and H. Campbell, 'Gene Environment Interaction: the BioBank UK Study', *Pharmacogenomics*, 2 (2002), pp. 75–82.
11. http://www.ukbiobank.ac.uk/ (accessed 22 May 2006).
12. http://www.generationscotland.org (accessed 22 May 2006).
13. R. Hapgood, 'The Potential and Limitations of Personalised Medicine in Primary Care', *British Journal of General Practice*, 53 (2003), pp. 915–16; G.C.M. Watt, 'What Will the New Genetic Information Do for Us?' *J Health Serv Res Policy*, 9 (2004), pp. 186–88.
14. B.H. Smith, A. Sheikh, G.C.M. Watt and H. Campbell, 'Genetic Epidemiology and Primary Care', *British Journal of General Practice*, 56 (2006), pp. 214–21; F.S. Collins, V.A. McKusick, 'Implications of the Human Genome Project for Medical Science', *Journal of the American Medical Association*, 285 (2001), pp. 540–44; Hapgood, 'The Potential and Limitations of Personalised Medicine in Primary Care', pp. 915–16.
15. J. Bell, 'The New Genetics in Clinical Practice', *British Medical Journal*, 316 (1998), pp. 618–20.
16. J.A. Knottnerus, 'Community Genetics and Community Medicine', *Family Practice*, 20 (2003), pp. 601–606.
17. T. Marteau and M. Richards, *The Troubled Helix: Social and Psychological Implications of the New Human Genetics* (Cambridge: Cambridge University Press, 1999).
18. D. Schrag, K.M. Kuntz and J.E. Garber, *et al.*, 'Life Expectancy Gains from Cancer Prevention Strategies for Women with Breast Cancer and BRCA1 or BRCA2 Mutations', *Journal of the American Medical Association*, 283 (2000), pp. 617–24.
19. Marteau and Richards, *The Troubled Helix*; S. Suther and P. Goodson, 'Barriers to the Provision of Genetic Services by Primary Care Physicians: A Systematic Review of the Literature', *Genetics in Medicine*, 5 (2003), pp. 70–76.
20. J. Emery, E. Watson, P. Rose and A. Andermann, 'A Systematic Review of the Literature Exploring the Role of Primary Care Genetic Services', *Family Practice*, 16 (1999), pp. 426–45.
21. E. Wallace, A. Hinds, H. Campbell, J. Horobin, R. Cetnarskyj and M. Porteous, 'A Cross-sectional Survey to Estimate the Prevalence of Family History of Colorectal, Breast and Ovarian Cancer in a General Practice Population', *British Journal of Cancer*, 91 (2004), pp. 1575–79.
22. K. Hunt, C. Davison, C. Emslie and G. Ford, 'Are Perceptions of a Family History of Heart Disease Related to Health-related Attitudes and Behaviour?' *Health Education Research*, 15 (2000), pp. 131–43.
23. Suther and Goodson, 'Barriers to the Provision of Genetic Services by Primary Care Physicians', pp. 70–76; Emery *et al.*, 'A Systematic Review

of the Literature Exploring the Role of Primary Care Genetic Services', pp. 426–45; A. Fry, H. Campbell, H. Gudmundsdottir, R. Rush, M. Porteous, D. Gorman and A. Cull, 'General Practitioners' Views on their Role in Cancer Genetics and Current Practice', *Family Practice*, 16 (1999), pp. 468–74.
24. S. Kumar and M. Gantley, 'Tensions Between Policy-makers and General Practitioners in Implementing the New Genetics: Grounded Theory Interview Study', *British Medical Journal*, 319 (1999), pp. 1410–13.
25. Last, *A Dictionary of Epidemiology*; Kumar and Gantley, 'Tensions Between Policy-makers and General Practitioners', pp. 1410–13.
26. Smith *et al.*, 'Genetic Epidemiology and Primary Care', pp. 214–21; Public Population Project in Genomics (P3G) Observatory: http://www.p3gobservatory.org/welcome.do (accessed 22 May 2006); M.A. Austin, S. Harding and C. McElroy, 'Genebanks: A Comparison of Eight Proposed International Genetic Databases', *Community Genetics*, 6 (2003), pp. 37–45.
27. Z. Chen, L. Lee and J. Chen, *et al.*, 'Cohort Profile: The Kadoorie Study of Chronic Disease in China (KSCDC)', *International Journal of Epidemiology*, 34 (2005), pp. 1243–49.

9

Genetics, Conversation and Conversion: A Discourse at the Interface of Molecular Biology and Christian Ethics

Brian Brock, Walter Doerfler, Hans Ulrich

Investigating genetics through collaborative conversation

This chapter reports on an experimental conversation between a practising molecular biologist and a Christian ethicist. It arose in the form of joint lectures in which the presentation of the technical state of the art in genetic science proceeded hand-in-hand with a theological analysis of the moral implications of its scientific models, discourses and hermeneutic claims. The impulse to open such dialogue was a sense from both sides that there is a serious deficit of detailed interaction between the two disciplines, creating a critical lack of relevant ethical discussion of issues related to human genetics. As a result, popular and academic discussions of ethical issues in human genetics have drifted apart to the point of absurdity. Yet rather than responding to this estrangement by embarking on the popular 'scientific education' approach, we felt that a concerted attempt was needed not simply to express the science to the public but to try to understand the moral implications of the science by struggling to articulate theologically expressed questions and criticisms in the course of discussion about the science.

The central focus of this discussion is human genetics. More specifically, we explored the relationship between the genetic knowledge of researchers, whose breakthroughs are so often publicly discussed, and the actual contemporary practice of medicine. Our aim is to elucidate the boundary between research on human biology and medical practice in order to clarify the interrelationships between the rapidly changing knowledge of researchers about precise aspects of human genetic functioning and the sweeping claims about treatment so often touted in the media with every discovery of a gene for this or that malady. Thus this chapter traces a dialogue between two voices in the broader discussion of medical ethics: the genetic researcher, whose work stands behind the promise of new future treatments, and the theological ethicist, whose aim is the clarification of the factors which sustain flourishing human health in its widest sense.

The main theme which emerged was that though theology and molecular biology are two quite different languages they are nevertheless critically concerned with the same subject matter. This chapter traces the points of agreement where we discovered the two discourses to converge on specific questions, though often the respective accounts of this convergence and the object of their shared investigation were formulated in strikingly different terms. The task was to find whether, and at which points, terms and descriptions in each discourse were flexible or rigid, and thus to identify previously hidden regions in which negotiation of perceptions were available. Both sides did, in fact, discover which aspects of their own positions were flexible and inflexible, and so clarified their own knowledge in the terms of their own discourses. Most importantly for the building of more nuanced discourse of genetic ethics in the future, however, this discussion allowed the language strategies of each side to emerge and so to become available for assessment and critique in ways not possible before the conversation. A new transparency began to develop regarding how language was being used to reveal or suppress aspects of the phenomenon on view through the use of specific metaphors or images. The result was a problematization of formerly assumed metaphorical constructs, and with this problematization a reformulation of familiar ethical questions became possible.

Because the whole experiment rests on the attempt to draw together two languages whose ethical implications at times point in different directions, it is important to state at the outset that the perspective of this chapter is theological. To admit this is not to imply that biological and theological languages are incompatible; rather, we should not too easily or quickly assume that they do not conflict, or that when they do the conflict is not without marked ethical implications. Such an unflinching search for conflict is sustained from a theological perspective by the Christian confession of the unity of all truth in Christ.[1] From such a perspective, exploring the conflicts between claims about creation is viewed in faith as promising for all participants in the discussion. This chapter attempts to find the basic points where ethically relevant conflicts between the two languages appear, in the hopes of founding an account of genetic ethics which is closer to the points at which scientific decision-making is taking place. We hope therefore that it might serve as a working resource for those at the forefront of scientific development.

Although in the day-to-day practical activity in both fields – molecular genetics and Christian ethics conflicts about ethical aims and procedures might appear unavoidable, what has emerged in discussion is that both disciplines in fact have the same goal: to understand, or at least try to understand, the universe in which humans have the privilege to live and the capacity to analyze. The rediscovery of this convergence of interest is a reminder of the Judaeo-Christian tradition that science is the study of the 'other book' of God's works, and as such is another expression of wonder at creation which seeks to know and understand it as divine gift. This definition also implies a basic limit to both scientific and theological enquiry. We cannot know the mind of God, which suggests that there will be types of cosmological questions which cannot be examined via the methods of science. Human study can, however, understand some of God's works. Here theological and biological enquiries share essential presuppositions. Both understand the primary place of attentiveness to phenomenon: in one case sacred texts, in the other physical processes. As a result, both have an acute sensitivity to the ways in which discourses become stale by straying from this close attentiveness into ideology by focusing, for instance, on abstract conceptualities such as systematization, or becoming too narrow exclusively deploying, for example, statistical analysis shorn of other experimental modalities. These insights framed our discussion and focused it on several points on the 'text' of human biological operation which is the sole subject of both biological science and ethical decision-making.

The first practical aim of discussion was to clarify precisely which aspects of the human genome were of shared ethical interest. Here, of course, molecular biology took a leading role in explaining what features of genetic processes were the focus of its investigations. Discussion of these critical biological junctures revealed that the deepest question for those interested in ethical questions is the extent to which interested parties are working with the same paradigm. Traditionally, the meeting-point for theology and medicine has been at the point of joint interest in the question of human suffering, and it is from this point that our discussion also found common ground. It is worth pointing out, however, that this essentially theological presupposition might not have been shared if one of the participants held alternative fundamental paradigms framed by primary ethical images derived from ideas such as evolutionary advance, genome enhancement or other foci. We were able to agree that what is given, created, in this case the human genome, contains and bounds human activities of study. At the same time we could also agree that this givenness, in exhibiting the richness of God's working, both serves as a limit to human activity and admits the reality of sickness and suffering. From such a starting-point theology and biology can share a complex view of human life in which sickness, as an aspect of human suffering, can be understood as a messenger telling us about biological life.[2]

Sickness is a part of life, and only by attempting to hear what it tells us about life are we able to understand the rich system of compensation and variability which is human life. Thus to ask about sickness brings the complexity of the given entity called 'human biological life' into view in a way which allows us to see how disease and healing are related in complex and intertwined ways which our interventions can only modify in relatively insignificant ways. This nuanced and interconnected relation of life and sickness is often the first casualty in popular discussion of genetics. By ignoring the complex relation of sickness and health, attention understandably falls to the exciting potential of gene modification, so placing a premium on that which is added or what is supposed to be newly created.

Our discussion avoided these oversimplifications by setting as its task the discovery of what the shared presupposition of the givenness of human life means when not assuming that 'human life' exists independently of our description of it. This awareness of the role of the subject in scientific investigation is the hallmark of modern science and theology, and grounds our agreement that our descriptions and investigations of life are already an answer to the givenness of creation and so already tied together with our hopes about what this givenness means. The discussion which follows is therefore not best understood not as a report on an example of 'discourse ethics' (which implies the reaching of agreement from fixed points of reference),[3] but as a joint exploration of biological phenomenon, which is 'prediscursive' in essence.

Reframing the basic presuppositions of genetic ethics

As already noted, contemporary discussions in medical ethics most often work with the background presupposition that the task of medicine is to minimize human suffering. In Western medicine the primary response to suffering is focused on the repair of biological defects. In the specialist medical ethics literature found in Western liberal polities this starting-point is then combined with economic or triage considerations to yield the sorts of ethical-limit discussions (e.g. plastic surgery and other elective surgeries, or the permissibility of sex-change operations). The more emotive of such questions often find their way into public discourse, as in the debates about therapeutic cloning, stem-cell research or gene-manipulated food. It is our contention that both popular and academic medical ethics discourses know too much: they already begin their thinking with a set of (false) assumptions about genetics. As a result they leave untouched the scientific progression in which the labelling of something as a defect carries an assessment that it requires a therapy. Taking into account the subjective framework of experimental science allows us to see the circularity of the claim that a biological defect is something which medicine ought to be able to remedy, a moving target which develops as concepts of health and sickness evolve.

This problem becomes particularly acute in the field of genetics. Consider for a moment what we have to assume to begin a research programme. For instance, we can be certain that scientists are not currently searching for a 'democracy gene' because the belief in democracy and capitalism are assumed by practitioners of modern science and are not considered a disease. Conversely, we have seen much sensationalized search for the 'gay', or 'God', or 'promiscuity', or 'crime' genes. That it seems sensible to research the genetic basis of homosexuality but not of democracy indicates how the genetic sources of various conditions are not only tied to human suffering but also to questions we can no longer talk about in public. Public funding is available to 'unravel' those questions that we as a society can no longer agree about naming pathological or normal. At these points the promise of genetics is to resolve debates which can no longer be resolved in public, research stepping in to offer clarity where democracy has broken down.

The intractable nature of ethical debates when framed in these terms reveals that, from a scientific standpoint, it is not at all obvious that we have been able to recognize what a biological defect really is. In relation to genetics, only in a limited number of instances is it even possible to identify what is properly thought of as a classical mutation. This complexity of the concept of mutation is rarely, if ever, questioned in discussions of ethics, with the result that the whole of the ethical debate in public and in the specialist literature has worked within the parameters defined by the question 'Which defects can and will we repair?' In the course of our discussions it became apparent that a great deal of additional analytical work will be required to provide a more adequate framework in which the medically oriented goals of providing therapies could hope to be realized. A knowledge-gap exists between genetic research, on the basis of which promises of cures are often presented, and the full development of genetic therapies. It appears that it may not have been well communicated to medical professionals and the political machinery which funds medical research that the state of knowledge in molecular biology is not sufficiently developed to be readily applied to the solution of important medical problems.

When the size of the gap between discovery and therapy is adequately appreciated, it becomes apparent that the common framing of discussions of specific questions in genetic ethics with the question 'should we or should we not' is premature. More importantly, this basic question is focused at the wrong point in the relation between research and medical treatment, namely, on the effects of therapies or of experiments which may never materialize. It is our contention that these questions of therapy and the ethics of its application are secondary or even tertiary considerations within a properly grounded ethical consideration of the issues, as they bypass the main question which an ethical consideration of genetic practices needs to answer.

Where the Western medical research and care projects meet Christian thought is in the desire to help the sick and to alleviate human suffering as much as possible. This desire explains the reasons why medicine might sometimes prematurely attempt to apply new concepts in molecular genetics. Nevertheless, it is counterproductive to make promises or raise hopes which cannot be realistically fulfilled.

In order to locate the progress of the state of the art more concretely in molecular biology, we may list some of the limitations of scientific knowledge of human genetic processes.

1. We lack full understanding of the regulatory mechanisms governing gene function.
2. We do not know the meaning of the great majority of the nucleotide sequence (repetitive DNA).
3. Epigenetic mechanisms (higher order information patterns) are only beginning to be unravelled.
4. Most medical problems have yet to be related to specific biological functions or to disturbances of function as affected by a range of genes.
5. Lack of function in one gene can sometimes be compensated, mitigated or exacerbated by the state of other genes.
6. Correlations between certain mutations and disease are statistically likely, but not always rigorously understood or proven.

From a functional point of view, the state of activity of a gene, or more likely sets of genes, is paramount to the understanding of the biological function and/or the causation of disease. We understand only partly how and why entire sets of genes are activated or silenced, the major alterations in activity profiles in cells or organisms which play a decisive role in embryonic and foetal development, in the causation of complex diseases, and particularly in tumour biology. A number of factors, frequently cumulatively termed 'epigenetic mechanisms', cooperate to effect the complicated regulation of activity profiles in sets of genes: the methylation of cytidine in DNA, modifications (acetylations and methylations) of DNA-binding proteins, the histones and other less well-understood modifications of chromatin. Evidence has recently been accumulating that the so-called repetitive DNA sequences, which comprise the majority of the human genome can participate in the overall activity regulation of the human genome. Not understanding their proper function means we are some way from understanding how disturbances in these unknown mechanisms might contribute to pathogenesis.[4]

Given the sweeping significance of the distance which scientific knowledge has yet to traverse before reaching knowledge which will reliably yield safe and effective therapies, Christian ethics raises the following question which must accompany not only the search for therapies but the explorative process itself. What, precisely, is a genetic defect? This question highlights two gaps in the presuppositions of the current scientific and ethical discussion.

First, neither scientific nor ethical discussions proceed with a sharp definition of the relationship between human sickness and genetic anomalies. In terms of the functioning of medicine, the concept of sickness plays a central role in deciding what a treatable illness might be. Thus, in order to talk meaningfully about the relationship of medicine and genetic science we must conceptually and empirically clarify how we ought to understand genetic changes to be related to human sickness.

Second, there is a much looser connection between genes and their expression than is typically assumed in public discussion. We are not yet in a position to *explain* what is going on between the genetic material of an individual and their phenotype, and are only observing hitherto invisible (because unsought) biological processes. We can observe relationships between symptoms and base-pair combinations, and attempt to describe them, but this observation, despite great progress, remains in its infancy. The term *gene expression* is at this point misleading as it implies a more direct and causal relationship than has yet been explained or substantiated.

Ethically relevant divergences of the languages of theology and molecular biology

The following three questions are trained on a single point: the establishment of the moral import of the biological phenomenon commonly known as the mutation. The basic insight expressed in these three intertwined observations is that genetic change is currently central to medical research and therefore must be the point at which Christian ethics makes its relevance apparent. We suggest three angles of approach to this problem. The first is to ask questions about the moral relevance of the heuristic nature of biological models. The second queries the moral implications of collectivist tendencies in biological models. Finally, we enquire into the moral relevance of language which describes changes in human genetic material – the so-called mutation.

1. A first problem complex lies in the use made in biological discourse of terms grounded on assumptions having non-scientific scope (e.g. relying on or implying theories about how human life began). This is a story, an image contained in metaphor, which has strong heuristic value. Terms like selection, speciation and survival are tied to an evolutionary metaphysic which has not been, and cannot be, scientifically substantiated by experiment. The usefulness of the terms is in marking out areas of exploration of phenomenon, and in shaping how these areas will be perceived. They thus function *heuristically* rather than *descriptively*. The heuristic role of evolution is easily visible in animal biology for example. Observing the difference in the depth of the fork in the tails of swifts, the evolutionary heuristic generates a testable first explanation. Different flight characteristics related to the different shapes might be keyed to different ecological niches and so provide different sorts of survival advantages. This theory of evolution does not determine whether this hypothesis is true or not, but is the condition for a highly specific sort of practical theorizing and investigation.

This heuristic power of the concept of mutation accounts for the focusing of current biological science on understanding the mutation. The evolutionary heuristic uses the term 'mutation' to designate the points at which biologically relevant or revealing events are taking place. Huntington's chorea can be understood to be functioning as a flagship example validating current explanations of genetic disease. A genetic change has been identified and linked with the occurrence of the disease. Here again, however, we must keep in mind that the mechanism of expression in even such a clearly correlated case is to date not understood. As so often, the genetic defect and the symptoms of the disease cannot be definitively linked. This illustrates how the scientific heuristic directs scientific scrutiny of phenomenon but does not itself produce results.

Having located the point of investigation, 'selection', 'survival' and 'speciation' name further theories about what is going on in a mutation, revealing and giving shape to investigations of specific occurrences in biological organisms. Yet these concepts have implications far beyond their heuristic value: they also carry strong *evaluative* overtones. It is the evaluative aspect of these terms which is ethically so important. While we can and must say that in certain cases something has changed in an organism, that the organism is functioning differently than expected, it is easy for the conceptual tools we used to reveal this facet of creaturely life to shape from the outset our moral assessment of the change. We may only look for something, for instance, we consider detrimental to the species. What allows us to see changes at all are evolutionary conceptual models, constantly under revision, but within which terms such as 'selection', 'speciation' and 'survival' play basic roles in anchoring the conceptual imagery.

Thus it is morally relevant to be clear that 'selection' includes the metaphor of improvement, of weeding out the less good, as does 'survival'. 'Speciation' asks about the borders and boundaries of genetic pools and thus locates specific cases within an enquiry of inclusion and exclusion. From a theological perspective, the task of separating the biological and political resonances of the theory are formidable indeed, and when not disentangled are immediately detrimental to the disabled.[5] If observed genetic change is labelled with one of these terms, a moral judgement is implied which frames future discussion about how such cases ought to be treated. We might rather say that in a modern age in which discovery and treatment are so closely intertwined, the aim of treatment, and the political measures it entails, are already assumed precisely in the search for and discovery of mutation.

This discussion of the moral implications of the theory of evolution revealed an important difference in how the theologians and the molecular biologist understood the status of the theory. The biologist among the authors of this chapter (Walter Doerfler) felt that when it was published (1859), Charles Darwin's *Origin of Species* may legitimately have been considered hypothetical, or speculative. But a century and a half of evidence-gathering, especially that provided by molecular biology's notation of the congruences between the nucleotide sequences of many different genomes, left no shred of doubt that all species have developed from precursors anteceding them in evolutionary time. His conclusion was that, although we still do not understand in detail the mechanisms that drive evolution, Darwin's main conclusions are correct. It therefore seemed to him unreasonable to question their overall validity, with the caveat that science has always remained open to unexpected and revolutionizing new developments and to the necessity to reconsider previously accepted concepts accordingly.

For the theologians the matter was less clearcut. They felt that their critical powers depended precisely on their ability to take evolution not as a truth-claim having equal validity to the Christian creeds and their claims about Christ, but rather as a provisional and developing human account of created reality.[6] Evolution is undeniably a theory of modern secular man, to which a Christian faith properly listens with eschatological patience, confessing in faith that this too is a partial truth, destined to be revealed and purified in the final judgement of human knowledge and activity. Theology thus cannot but see evolution as part of the modern tradition of scientific enquiry, a tradition which was nurtured and is still supported by many Christians. Because evolution is a central belief of the modern scientific tradition, comments Oliver O'Donovan, it is not easily held as provisional by its participants. '[O]ne cannot honestly relate to one's tradition like the conservator of a museum. It has to be lived in confidently.'[7] Tradition makes common action possible, meaning that the community of scientific enquiry exists only as it agrees on a truth, in this case the truth of the theory of evolution. Because the Church has not yet seen fit to add the theory of evolution to the list of things it confesses to be finally and undeniably true about reality, there remains at this point an eschatological tension between the tradition of faith, and the tradition of scientific enquiry. This does not mean that Christians need deny the theory of evolution, but that their membership in a community anchored in claims about Christ gives them a freedom to remind the community of scientists that its truth-claims are of a different order than its own fundamental truth-claims. That reminder is that modern science has based its authority on the claim that all truly scientific claims take the form of statements which might be, but have not yet been, falsified.

2. This raises question of the *location of genetic sickness*. An obvious initial question to be asked is whether sickness is to be understood as something attributable only to the individual or to the individual *and* to the whole. In scientific terms this is a question about whether to refer to the sick according to phenotypic or genotypic attributes. Is someone sick when they carry a gene or when they exhibit a pathology?

In Christian theology, only an individual, an 'I', can be sick.[8] To say this is to reveal that the claim that the 'whole', the genome, is sick can only be a metaphysical or theological claim. Such a concept of a 'sick genome' is a pessimistic stance about the human race and amounts to a questioning of the viability of the whole biological system. The scientist, therefore, must leave this question to one side. In order to make any claims of a scientific nature about the viability of the gene pool, its 'sickness' would require a control group. This would require another human gene pool to measure deviations in our gene pool against – an obvious impossibility. Science, on its own terms, is left with only this biological system called the human race to study in as much detail as possible.

The Christian response to this problematic is to affirm that whether the human genome is slowly changing or not, it is the biological material which we have been given, and so is good. In a theological sense, to exist at all is good: an artefact of God's provision precisely in a fallen state. The 'And it was good' of the first two chapters of Genesis ought not be understood as having been revoked because of the Fall. The implication of this Christian affirmation is the claim that sickness can only be a statement about the suffering of the individual. The gene pool cannot be sick.

The ethical result of putting this so strongly is an insistence that sickness must be so defined that the sick individual remains part of the human community in all circumstances. They are not excluded from the community, excommunicated, by possessing a divergent genetic heritage. No matter their genotype they remain one of us, claiming our medical care for the alleviation of their particular sufferings rather than using them as an example of a divergent or threatening stream within the gene pool, to be cut off, excluded or studied for the good of the whole.

To make such a firm distinction exposes the fact that the scientific heuristic often works against it. Biological definitions of sickness arise from heuristic apparatuses which have no way to make precise distinction between the sickness of the individual and the sickness of the whole. If disease is considered part of the functioning of the whole biological system then the sickness of the individual is *per definitionem* a marker of lack of, or part of, the process of the adaptation of the whole, reducing the priority of treating the individual *qua* individual. This tendency of the heuristic of biological science has clear ethical implications for how we approach sick individuals both as individuals and as members of society. Medicine cannot and must not treat society or gene pools. The political task is to guard against the force of the metaphor toward thinking in terms of collectives. Special vigilance in oversight is therefore appropriate when methods which explicitly work with collectives play a dominant role in policy-making, such as genetic epidemiology[9] and public health.

3. These observations take us to the deep ambiguity behind these problems: what is a *mutation*?[10] Unlike sickness, which we discover from the cries of the subject who suffers it, the language of mutation assumes an observer's perspective. To pronounce something a mutation requires a view of the whole gene pool and a theory of its development. We have already seen that, empirically, this is information that scientific method cannot supply. All that might be said about the constitution of the whole gene pool is therefore hypothetical, extrapolated from the major part of the human nucleotide sequence which is now available. On the basis of the information about the genome sequence possessed to date, researchers can compare sequences from any individual. Yet to so define normal function relies not on comprehensive, or even broad, knowledge of the content of the whole actual genome, but only a statistical 'normal' extracted from a minute portion of it. Here the popular force of Darwin's account often misleads. In *The Origin of Species*, Darwin's most memorable examples of special variations in gene pools draw on observations about the breeding of plants, pigeons and dogs.[11] But breeders *select* for an outcome, an experimental procedure not available to scientists. The problem faced by scientists is that of the human life-span: what can be empirically established without the verifiability test of breeding?

Biologically, we can speak more specifically about the meaning of the term mutation. Genetic information in any genome can be altered, mutated, in several ways. There can be single nucleotide exchanges; deletions, insertions of single or multiple nucleotides; sequences can be amplified; mutations can be located inside genes or may occur in the regulatory region, or they can affect the reading frame or the splicing mode of a gene. Gene expression can, moreover, be subject to epigenetic mechanisms; among them the sequence-specific introduction of methylated nucleotides and the modifications of histones have a decisive effect. In a number of instances, we can directly explain in definite biochemical terms how a mutation alters the function(s) of a gene (e.g. in sickle cell anaemia or in cystic fibrosis), and we have a good idea how diseases arise as a consequence of these mutations. In a much larger number of cases, there exist only correlations between mutations in a given gene and a disease, while we do not possess the absolute proof for the validity of our correlative conclusions. Moreover, in complex genetic alterations, as in chromosomal aberrations, even when the genetic change is well understood, its mechanisms are not. For instance, with Down's syndrome, we lack an explanation of how the presence of a third chromosome 21 can lead to the phenotype of Down's syndrome, even though the complete nucleotide sequence of some 33×10^6 and most of the genes it carries are well known.

The relevant point in the context of our discussion is the ways in which mutation language is, at least in one of its most important resonances, anti-theological. At one level we can allow that 'mutation' names the various alterations just discussed. In some cases the changes are necessary and regularly occurring in every organism; in other cases familiar biological processes are hampered. It is this harmful version of change which we commonly think of as a mutation. Analysis of the cultural freight of the term 'mutation' thus reveals that it also entails an *assessment* of the change, that it is a harmful change. 'Mutation' thus often assumes a judgement about the 'survivability' or 'adaptability' of an organism defined by reference to a specific point in its genome. The moral assessment built into the term 'mutation' thus calls forth a judgement about whether the change represents a benefit or deficit, and so goes far beyond simply noting a change in the expected sequence of nucleotides.

It is this *judgement* of change which draws biological science beyond a description of biological processes and into a cosmological theory with ethical import. As such it is not 'another perspective' alongside the theological, but a counter-theology which takes the position of the last judge in making quasi-eschatological judgements about the whole genome and its trajectory. It is here that the tight modern linkage of therapy and discovery reveals its problematic aspect. The language of mutation oscillates between a scientific description and a medical diagnosis which implicitly calls for treatment. Mutation carries a negative medical status which drives the impulse to repair. But not all changes in the human genome are so easily labelled 'mutations' in the sense of 'defects to be repaired to alleviate suffering'. One example of this might be Down's syndrome. Yet keeping separate the descriptive and prescriptive, the explorative and applicative, is especially problematic in the context of the modern interlinkage of science and technology in which politicians supplying the vast sums spent on medical research demand at least the promise of therapeutic or economic benefit.[12]

The richness and variety of the gene pool must not be equated with the richness and variety of human flourishing because this, in theological terms, is a richness of relationship. Mutation language is built on a picture of the watchmaker God, who created and leaves creation alone. God is thus alienated from his creative work, and we are severed from understanding ourselves as part of God's ongoing work. In Christian theology, biological change is part of a larger story, one of the many subnarratives within a larger narrative of God's working, the end of which we do not yet see. We have no perspective from which to draw a definitive conclusion as to whether a biological change is part of the new creation or not. We therefore cannot say 'this is superfluous genetic material' or 'this is a defect, a mutation' unless we take up a totalitarian perspective in which we do not just claim to know parts of the logic of the whole, but that we exhaustively know the logic of the universe.[13]

Conclusions

There are two main conclusions which have come out of the biological–theological dialogue that has formed this chapter. The first of these is that biological and theological discourses stand before the same basic material processes, but are equipped to emphasize and question different aspects of it. It is in examining the phenomenon itself that practitioners of both disciplines meet. This meeting-place supplies a criterion for defining what it is for both disciplines to be practised well. This criterion is attention to phenomenon. If both disciplines are understood this way we are enabled to see that they utilize similar and intertwined modes of enquiry. As they face the 'given to be described', the two wrestle to define what is being described in the light of the ethical question, 'What are we to hope for?', the combinations of these questions yields the heuristic question which can properly guide scientific and theological research: 'What, precisely, is the relationship of the given to that which we hope for?'

Here, to say that we 'understand' is not simply to make a claim correct in analytical or linguistic terms, but is measured against the multiform standard of phenomenon. We must seek to understand what is going on *here*, with this specific biological process in all its interlinked complexities. This has led us to stress in this chapter not so much the two languages which we bring to the enquiry into biological processes, but rather the search to find how the two logics of enquiry embedded in the disciplines of biology and theology might be understood to meet in the quest to understand the suffering of humans. In so far as they do, then they can be understood in a newly recovered fashion to be a joint enquiry.

Our joint inquiry into the role of genes in human life has led to the discovery that it is not the case that all scientific research into the human genome is essentially oriented by the hope of relief of human suffering. At the same time, a strong overlap between some biologists and theologians emerged in their shared presupposition that what is under study is not properly understood through a technological desire to create something new or 'improved' but can only be properly grasped if what currently exists is taken on its own terms, without any desire to change or improve it. This shared presupposition about the givenness of biological systems can also meet a shared hope for the future of humanity in which suffering is ameliorated. In so far as theologians and biologists share these presuppositions and aims, the descriptive and therapeutic remain intertwined in a manner allowing fruitful enquiry into the moral relevance of our descriptions of phenomenon. Because modern research into genomics takes place within a modern context in which research projects are closely tied to application, it has been our claim that genetic ethics must do its most important work at this fundamental level. Only by understanding the moral implications of the way our naming of phenomena is already loaded with ethical implications do we actually deal with the ethical questions of our age.[14]

The second conclusion is that to facilitate this wrestling by theologians, a productive grasp of biological phenomenon is necessary in order that a greater awareness may develop of the points at which biology is asking its questions and of the decisions currently facing biological scientists. For instance, discussions of therapeutic cloning depend on the diagnosis–sickness–repair model, the ethical implications of which we have queried. In order for such questions to be properly and accurately understood, our enquiry forces the opening question of precisely how 'therapy' is being, and ought to be, defined. Much of the work of theological ethicists on the topic of genetics simply grants that 'science' has defined the meaning of phenomenon, and so simply gives away its ability to speak with moral relevance through lack of proper engagement with scientist and the phenomenon they observe.

At the same time biologists must become more reflexive about what they do. They must come to see the far-reaching implications of the claim that their work not only discovers what exists in the world but 'imposes and structures' rather than simply 'discovers' and 'repairs'. In being asked to speak publicly with theologians about *how* they come to understand biological phenomenon, biologists become aware of the ways that their images and metaphors impose and structure perception in the course of research. As they do so, they may become aware of the way in which their work is driven by powerful hermeneutic filters in the form of insightful languages and metaphors which both make their work possible and also carry strong moral implications. We have outlined the points at which the metaphors which allow biology to see also push in the direction of moral claims which they may not wish to endorse. It is the task of theological enquiry to play the role of midwife in revealing the importance of these morally critical points. When the biologist undertakes a public description of his or her experimental approach, they are given opportunity to see with more clarity the richness of the hermeneutic that makes perception possible, while at the same time offering access to the moral implications which their own hermeneutics press on them.

We have seen the blunt polarities almost always assumed in moral debate about genetics, such as 'genetic vs environmental causes', 'normal vs mutated gene' and 'coding vs extra repetitive genetic material', obscure rather than clarify the moral questions at stake. Genetic scientists too can show us that much remains unclear, and all interpretative. The promise of genetics to provide moral clarity and mastery of phenomenon has, on further inspection, been shown not only to be less clearcut than we once believed, but the quest for this clarity has been exposed as an evasion of the complexities of politics and moral deliberation about who humans are and what they are to hope for.

Notes

1. For a recent treatment of this theme, see Gavin D'Costa, *Theology in the Public Square: Church, Academy and Nation* (Oxford: Basil Blackwell, 2005), ch. 6. D'Costa draws on the long Catholic tradition of interdisciplinary dialogue which is summarized by John Paul II: 'Science can purify religion from error and superstition; religion can purify science

from idolatry and false absolutes' (p. 198).
2. Cf. Michel Foucault, *The Birth of the Clinic: An Archaeology of Medical Perception* (trans. A.M. Sheridan Smith; New York: Vintage, 1994), chs 8–9.
3. In the form popularized by Jürgen Habermas. See *The Theory of Communicative Action*, vols 1 and 2 (Boston, MD: Beacon Press, 1984 and 1987).
4. For more detailed discussion of these points, especially as relating to DNA methylation, see, Walter Doerfler and Petra Boehm (eds), 'DNA Methylation: Basic Mechanisms', in *Current Topics in Microbiology and Immunology*, 301 (2006), and Walter Doerfler and Petra Boehm (eds), 'DNA Methylation: Development, Genetic Disease and Cancer', in *Current Topics in Microbiology and Immunology*, 310 (2006).
5. Consider the following comment from Charles Darwin, *The Descent of Man and Selection in Relation to Sex* (London: John Murray, 1906):

> With savages, the weak in body or mind are soon eliminated; and those that survive commonly exhibit a vigorous state of health. We civilised men, on the other hand, do our utmost to check the process of elimination; we build asylums for the imbecile, the maimed, and the sick; we institute poor-laws; and our medical men exert their utmost skill to save the life of every one to the last moment. There is reason to believe that vaccination has preserved thousands, who from a weak constitution would formerly have succumbed to small-pox. Thus the weak members of civilised societies propagate their kind. No one who has attended to the breeding of domestic animals will doubt that this must be highly injurious to the race of man. It is surprising how soon a want of care, or care wrongly directed, leads to the degeneration of a domestic race; but excepting in the case of man himself, hardly any one is so ignorant as to allow his worst animals to breed. (pp. 205–206)

6. For discussion of this point, cf. Karl Barth, *Church Dogmatics III.3* (Edinburgh: T&T Clark, 1960), pp. 124–31:

> No law is known to us with the certainty with which God is known to us by His Word, or with such clarity that even in relation to its own sphere we can responsibly pass it off as at least the formal foreordination of all actual occurrence within that sphere. No high measure of noetic certainty or clarity can give to laws known to us, i.e., discovered and guaranteed by us, the character of ontic laws, the character in virtue of which we necessarily perceive in them the laws of God, and therefore in effect the real foreordination of creaturely occurrence at any rate from the formal standpoint (p. 127).

7. Oliver O'Donovan, *Common Objects of Love: Moral Reflection and the Shaping of Community* (Grand Rapids, MI: Eerdmans, 2002), p. 38.
8. Such a statement is the fruit of a long and rich Western discussion of the meaning of sickness, recoverable due to the revitalization of

the philosophical *principium individuationis* in the wake of Friedrich Nietzsche's work. Hans Günter Ulrich, *Anthropologie und Ethik bei Friedrich Nietzsche: Interpretationen zu Grundproblemen theologisher Ethik* (Munich: Christian Kaiser, 1975), pp. 61–66.
9. Discussed in Blair Smith's contribution to this volume (Chapter 8).
10. The oscillation of perception marked by the terms 'variation' and 'mutation' lies at the heart of the problematic of Darwinian theory, present from its inception:

> The term 'variety' is almost equally difficult to define; but here community of descent is almost universally implied, though it can rarely be proved. We have also what are called monstrosities; but they graduate into varieties. By monstrosity I presume is meant some considerable deviation of structure in one part, either injurious to or not useful to the species and not generally propagate. (Charles Darwin, *The Origin of Species* [Oxford: Oxford University Press, 1996], ch. 2, p. 38)

Darwin's main theoretical insight was that the difference between a variation and a mutation is less straightforward, and more interpretative, than often assumed.
11. Darwin, *The Origin of Species*, ch. 1.
12. For a related account of the intertwining of modern experimental science with technology, and the political forces that funding brings to bear on the direction of research, see Mary Joe Nye, *Before Big Science: The Pursuit of Modern Chemistry and Physics 1800–1940* (Cambridge, MA: Harvard University Press, 1996).
13. Bernhard Waldenfels, 'Ordnung im Potentialis: zur Krisis der europäischen Moderne', *Der Stachel des Fremden* (Frankfurt am Main: Suhrkamp, 1998), ch. 1, pp. 15–27.
14. To our knowledge, the only medical ethicist making similar arguments is Johannes Fischer, Professor of Theological Ethics and head of the Institute of Social Ethics at the University of Zurich. See 'Über die schwierige Beziehung zwischen Forschung und Ethik', at http://www.ethik.unizh.ch/ise/downloads/publikationen/fischer/Fischer-Forschung_u_Ethik.pdf.

Part 4

Theological Reflections on the New Genetics

10

Life's Goodness:
On Disability, Genetics, and 'Choice'

Hans S. Reinders

The bioethical literature on moral issues regarding genetics – prenatal testing for genetic disorders, genetic screening, selective abortion, eugenics, to mention just a few – generally presupposes that a genetic disorder constitutes a 'defect', which explains why most authors in bioethics share a presumption in favour of 'cure', and, or, when this is impossible, 'prevention'. With regard to this presumption the bioethical literature faces a problem, which is that quite frequently people with genetic disorders resent the notion that their lives are somehow 'defective'. By implication, they also oppose the claim that it would be better to have them changed, or, if impossible, to prevent them from coming into existence altogether. Bioethicists have responded that this opposition is misguided in so far as it assumes that the use of clinical genetics is premised by a negative evaluation of their lives. It is not the case. It aims at enhancing people's freedom to make responsible decisions regarding their offspring in their light of their own values. Elsewhere I have dealt extensively with the questions involved in this controversy, including questions about its political ramifications in liberal society, which allows me to devote my space here to a different issue.[1] It is presented by the fact that many people with disabilities resulting from genetic disorders experience their own lives as good.

While not denying the legitimacy of reproductive choice, some of the authors writing on bioethics and disability maintain that people with disabilities perceive the implicit message of clinical genetics to be that they are living less valuable lives.[2] In philosophical terms: the underlying issue of the controversy on genetics and disability is about the good life for human beings. In this chapter I will interpret the critique of the devaluation of disabled lives in a strong sense, namely as saying that life is good as it is. This is not to deny, of course, that there are many goals or ambitions to improve such lives, which is generally true of most human lives. It is to deny, however, that the condition of being disabled, or of sharing one's life with a disabled person, is *as such* detrimental to life's goodness. On the contrary, people with disabilities frequently express appreciation not only for being alive, but also for living their kind of life, indicating that the social responses to their condition are informed by prejudice more than anything else.[3] What is particularly interesting about their appreciation is that it is acquired in the process of learning to live a life that most disabled people would never have chosen for themselves, had there been a choice.[4] The topic of my reflections, therefore, is to try to understand what it means to say that life is good as it is, not because we chose it, or would have chosen it had there been a choice, but simply because it is what it is.

Given the present context, the focus will be on disability, intellectual disability, to be more precise. In particular with regard to intellectual disability – which in my usage is roughly the same as 'mental retardation', 'developmental disability', 'cognitive disability', or 'learning disability' – the experience of life's goodness appears as doubtful, given the fact that living with this condition is generally considered in a negative light, often even within the disability community itself.[5] This fact leads me to suppose that the most important task of ethics with regard to intellectual disability is to reflect upon what it is to appreciate life as it turns out to be. The question I am interested in is, therefore, what is presupposed in the notion that life is good as it is, even when it is not the 'normal' kind of life that people in our moral culture seem to think they are entitled to.

Since this introduces a type of reflection that is quite distinct from what is usually offered in textbooks on the subject of ethics and genetics, I will start with some of the assumptions that need to be clarified in order for the reader to understand where I am coming from and where I am heading in this essay. Once the stage is set, I will begin to explore the differences between two rival conceptions of the human good; one that makes life's goodness dependent upon 'choice', the other that makes life's goodness dependent upon 'gift'. The aim of this chapter must be quite modest, given the many questions it invokes. I hope to establish how listening to accounts of people who experience their lives as good in spite or because of a condition of disability may inform a different type of ethical reflection. For reasons to become apparent later on, I will switch from philosophy to theology in the course of the argument.

Setting the stage

In view of the many questions arising in the context of contemporary medicine, bioethicists often presuppose a particular task of ethical reflection, which is to establish and defend a taxonomy that enables moral judgement. There are things people ought to do, there are things they never should do, and there are things they are allowed to do. It is the business of ethics to sort out the differences and draw the lines between them. Ethical reflection, this is to say, is mapping the areas of the morally right, the morally wrong and the morally permissible. Normative ethical theories provide the substance for these distinctions, which can then be applied to particular areas of moral concern. This is the textbook account of what bioethics as an instance of applied ethics is about.[6]

When it comes to the discussion on genetics and disability, this conception of ethical reflection informs discussions about moral constraints on medical interventions, on acquiring and using information, or on trying to eliminate certain genetic disorders, etc. As indicated, however, I doubt that the real issue is about distinguishing the morally right from the morally wrong, or the morally permissible. Instead, I want to suggest, it is about what we regard as the good of human being. The reasons why people think that disabled lives should be cured, or should be prevented from coming into existence have to do with what they regard as a worthwhile life. The condition of being disabled, particularly of being intellectually disabled, is hardly ever part of what people believe about a worthwhile life. On the contrary, this condition is among what most people think should be avoided when it comes to having a good life. To lend plausibility to this claim, I will present five assumptions about the debate on genetics and disability in liberal society.

The first assumption has to do with using 'genetics' as a way to prevent disability. I take the general attitude towards genetics in this respect to be one of moral ambivalence. Evidently there is a lot to be gained for human beings when the causes of certain diseases can be traced back to their most elementary components – faulty genes producing the wrong kinds of proteins at the wrong place or the wrong time. Whole families need no longer live in fear that their offspring will be affected by them. Early detection of genetic disorders appears therefore as basically a good thing because it can inform people about their prospects of having a normal and healthy life with their family. On the other hand, however, there are many worries about where the explosive development in genetics in recent times may take us. The possible range of medical interventions in the fundamental conditions of life is such that scientists seem to be 'playing God', as it is sometimes put. Many people perceive the combination of genetics with artificial reproduction as the road towards choosing the 'perfect baby'. The science of genetics seems to have revived the old Promethean dream of humans being in control of life itself.

Geneticists and bioethicists usually do not have much patience for this kind of worry, which they claim is largely exaggerated.[7] The possibilities of genetics are more limited than people usually tend to believe, particularly in respect of the issue of eugenics. Contrary to what many authors in the field of bioethics have argued, however, there is no clear distinction between therapeutic uses of genetics in order to prevent human suffering on the one hand and enhancement of our offspring by choosing particular kinds of children on the other. Not only does our society allow us to decide how many children we will have, and when and by whom, but it also allows us to decide what kind of children we want, and – by implication – what kind of children we do not want. In the latter type of decisions the lives of people with disability are at least implicated because the interventions of clinical genetics could not make sense without being intended to prevent certain kinds of lives. Even when this is not the explicit aim of clinical genetics, it is nonetheless its unintended side-effect because eugenics in liberal society is consumer-driven. Given the fact that our society accepts that we are entitled to our own conceptions of the good, there is no limit – as a matter of moral principle, that is to say – to the range of biological conditions that may be targeted as the appropriate object of genetic manipulation as long as our aspirations to do so do not violate other people's legitimate interests. By the same token, there is no limit to the range of biological conditions that can be targeted as a 'disability'. Any attempt to impose limits to medical interventions of a certain kind – for example, interventions qualified as 'genetic enhancement' – will sooner or later be exposed as morally arbitrary. This is my first assumption.

The second assumption regards what is commonly taken to be the role of ethics in all this. The attitude of ambivalence implies that people see 'pros' and 'cons' with regard to the uses of clinical genetics, which to sort out is generally perceived to be the task of ethics. Given this perception, the logical question to ask is where to draw the line. For example, which possibilities of prenatal testing combined with artificial reproductive techniques deserve our moral support and which don't? As indicated, I interpret this question as the question for a mapping of moral distinctions. Frequently, the taxonomy involved invokes the metaphor of a 'grey area'. This metaphor suggests legitimate and illegitimate moral spaces, as well as spaces where the distinction is unclear. If genetic testing can prevent a life with the syndrome of Lesch Nyhan, to use a well-known example, people tend to think this a good thing, given what is known about the suffering that this syndrome involves. If the same technology is used to prevent a life with Down's syndrome, however, the justification seems to be more problematic, given the many happy lives that are lived by people with that condition.

Many arguments in the bioethical literature about these examples show that people are divided on where boundaries should be drawn, but this is presently not the point I am interested in. The point is that the task of ethics is regarded to be the identification of these boundaries. Ethics should draw distinctions and lay out the moral landscape and indicate the areas where the new genetics may take us as well as the territory it should avoid. The present debate on stem-cell research is just another example of the disputes on moral mapping. This, then, is my second assumption. It says that the task of ethics is to set moral boundaries.

The third assumption regards a question of moral substance. What is it that ethical reflection provides in terms of the moral taxonomy that is expected from it? The dominant approach in contemporary bioethics answers by producing rules and procedures in order to put moral constraints on human action. This approach is governed by the presumption of individual freedom. People have a right to find out and prevent any genetic condition they want as long as certain moral requirements are taken into consideration. Questions are raised about the right to know, or the right not to know, for example. Other questions regard the right to protection from illegitimate uses of genetic information constituting genetic discrimination. All of these questions regard the distribution of the rights and wrongs of usages of genetics in a medical context given the various interests that are involved.

Take as an example the requirement of paying respect to people's privacy. If someone wants to find out the risk of being disposed for a particular disease, the result may have serious consequences for their family members, which indicates that there may be a moral constraint on finding out about one's genetic dispositions for reasons of protecting other people's privacy. Another example is the constraint on using this information once it is obtained. If a person is known to have a genetic disposition there may be other parties that are interested in this kind of information, insurance companies for example, or employers, which indicates another issue of privacy. Characteristic for this approach to ethical reflection, then, is a moral taxonomy that is shaped by the freedom to acquire any kind of genetic information and act upon it, provided that moral constraints are respected in the use of this information by, or with regard to, other people.

The fourth assumption regards the question that this approach to ethical reflection leaves out of account, which is the question as to why people turn to the possibility of obtaining information about their genetic dispositions in the first place. Usually, people seek genetic information when they belong to a group of people with an increased risk for certain disorders or dispositions either in themselves or in their children, or when there is a suspicion of hereditary disease in their family which they do not want to pass on to the next generation. Not all of these disorders and dispositions result in diseases that seriously affect the quality of life of those living with them, but some do. I already used the examples of the syndrome of Lesch Nyhan and of Down's syndrome. The contested matter here is 'quality of life'. Some people believe that there are disabling conditions that are compatible with living a rewarding human life, while most others think such lives to be horrific. The underlying issue in these kinds of disputes regards the conception of the good that is guiding our moral judgements. Decisions regarding the consequences of genetic dispositions are guided by a conception of the kind of life people want to live and, by implication, also by a conception of the kind of life they want to avoid. The latter is necessarily derived from what they know – or think to know – of the lives of people living with the same disposition. The kind of life they want to avoid is lived by certain people. This indicates how the lives of people actually living with genetic disorders or dispositions are negatively implicated in decisions to prevent their kind of life. Without a conception of what these lives are like, there is no point in making such decisions. Accordingly, the possibility to cure or prevent disabled lives renders such lives as proper objects of medical intervention. In any given case, this rendering implies that these lives are different from what they are expected to be, *qua good human lives*, this is to say. Therefore, to argue for or against moral constraints regarding the object of 'therapeutic intervention' is to argue for or against the implicit conceptions of the human good that make sense of these interventions. This is my fourth assumption.

My fifth assumption, finally, regards the political context within which decisions about the consequences of one's genetic dispositions are made. Liberal society strongly believes in bracketing issues about the good life from moral argument because of the value it places on individual freedom. It believes that people should be free to live the kind of lives they want for themselves, according to their own conception of the good, without being bothered by the fact that other people pursue other conceptions. The bioethical literature reflects this moral belief as one of its basic principles. It is morally all right for some people to think that there is nothing wrong with living a disabled life as long as they respect other people's right to disagree. Consequently, when developments in genetics and medicine provide people with a choice about what kind of children they want or do not want, their options cannot be curtailed just because of what other people think about the decision they make. The fact that you think that your kind of life is all right in no way commits me to thinking the same, let alone that I could be legitimately forced into living your kind of life.

If one accepts the foregoing assumptions then the argument for my claim with regard to the task of ethical reflection on genetics and disability is readily made. Setting moral side constraints on how people use information about their genetic dispositions is guided by what moral philosophers used to call 'other-regardingness'. Decisions about the consequences of genetic dispositions are judged to be right, or wrong, or permissible, this is to say, in respect of other people's moral interests. They are not so judged from the perspective of their 'self-regardingness'. What our decisions do to our own interests is not other people's concern, and, therefore, not considered to be the proper object of moral concern. The reason is that our society believes that where our own interests are concerned we should be allowed to make decisions guided by the light of our own conception of what we take these interests to be, irrespective of what other people may think about them. The other-regardingness of moral concern in liberal society reflects what Alasdair Macintyre has called the 'privatization of the good'. In the present context, this means that decisions regarding the consequences of genetic dispositions are accepted without question to the extent they regard our 'private' lives. They are made in the private domain of the people whose decisions they are.

The guiding principle of this approach to ethical reflection on genetics and disability is in fact the principle of individual freedom in deciding one's own conception of the good and act upon it with equal respect for other people's freedom. This principle supports the task of ethics as it is commonly understood in liberal society. Ethical reflection is supposed to stay away from conceptions of the good because they are not supposed to be disputed as part of our moral discourse. With regard to the moral issues raised about the interventions in clinical genetics, this view implies that a decisive question remains outside the scope of moral discourse. This is the question of whether a genetic 'defect' makes a human life 'defective' in respect of its goodness. Many people in our society do believe that it does. Many people with disabilities and their advocates disagree. They have learned to appreciate their lives with all that it brings.

Having a life and being alive

Living a life with a disability, or sharing one's life with someone with a disability, can be a rich experience, not unlike most other human lives.[8] Parents of disabled children who make this kind of claim, however, are sometimes answered with a question: 'Were you in a position to choose whether to have this child or another, healthy one, would you choose the same child again?' Presumably, this question is intended to expose a dilemma facing these parents. If they say they would choose the healthy child, their lives can not be as rewarding as they suggest. If on the other hand they say they would choose the same child again, however, their choice appears less than credible. What reason can one possibly have for desiring a disabled child?

What should not go unnoticed, however, is the presupposition of this query that affirmation by choice is apparently what makes our lives 'really' good. When parents say that they are experiencing their life with their disabled child as rewarding, it is the condition of its being chosen, hypothetically at least, that supposedly confirms their judgement. Without that kind of confirmation, their judgement is perceived as doubtful, to say the least. Their life cannot really be good, that is to say, unless they would say that they would choose the same life again. This sceptical attitude indicates a conception of the good that I characterized earlier as dependent upon choice. It is our choosing it that apparently confirms the proclaimed goodness of what we choose.

These observations indicate a view within which the point of having a human life is in the condition that it is at our disposition. To have a life is to be its governor in the sense that the more of its conditions are under control the more we have reason to affirm its goodness. It is to be able to say, 'Yes, if I had a choice, I would have the same life again'. Not only the affirmation but also the negation answers to the same condition: 'If I could do my life over again, I would do it differently'. To have a life in this dispositional sense implies having the desired span of control over its conditions. In other words, life's goodness and 'choice' are, at least hypothetically, connected.

Evidently, the preoccupation with 'choice' has no place in the notion of human life that is good as it is. Life's goodness, according to this notion, does not depend upon 'choice', not even hypothetically. It is not that life as it is is good because I would choose the same life again, if there were a choice. If my life were different from what it happens to be, then it would also be good. Life's goodness, this is to say, does not relate to 'choice' but relates to 'being'.

To grasp this distinction, it may be helpful to look at how it has been presented in connection with 'end-of-life issues' in bioethics. There it appeared as the distinction between 'having a life' and 'being alive'.[9] This distinction has been proposed to identify the condition that makes a human life worthy of moral esteem. It points to the fact that in some cases human beings are 'merely' alive, without having a sense of their own existence. They exist biologically, but not autobiographically. 'Being alive' one cannot experience one's life as one's own. One cannot de said, literally, to own one's life. Life is only properly one's own if one has a life. This explains why bioethicists have used this distinction to justify that people who live in a permanent state of unconsciousness cannot be said to have a human life properly so called, and that, therefore, their lives cannot have the moral standing that human lives in the full sense of the term have.[10] Consequently, some of them have argued, ending these lives cannot be illegitimate, other things being equal, because one cannot violate the legitimate interest that human beings have in their lives when they cannot experience their lives as their own.[11]

Now, if we contrast the conception of having a human life in such arguments with the notion of 'being alive', it is immediately clear why the latter must fall short of the former. When it is said of me that I am alive, my being appears differently from when it is said of me that I have a life. 'Being alive' simply 'is'. As a matter of fact the term 'alive' does not seem to add anything in particular. It lacks any reference to disposition. No arrangement, no plan or project, no expectation is entailed in the notion of being alive.

Having a life, in contrast, seems to imply a different kind of being. Beings of whom we say that they are capable of having a life meet the condition of a being capable of evaluation. Beings of which we say that they are alive do not necessarily meet this condition. Of some human beings it is said that they are 'merely alive', human beings in a persistent vegetative state, for example. In our contemporary culture this statement usually has the meaning of a verdict. The humanity of those of whom we say that they are 'merely alive' is a matter of dispute, to say the least, whereas this is not the case with regard to those that have a life in the dispositional sense.

When we return to the view that makes life's goodness dependent upon choice, either actual or hypothetical, it immediately follows that those of whom it is said that they are 'merely alive' cannot have a good life. This is confirmed by how liberal authors in bioethics conceive of the value of human life. From their point of view persons with profound intellectual disabilities, for example, cannot possibly have a good life. They have no plans, no expectations, not even complaints; they are 'merely alive', which explains why living a profoundly disabled human life is sometimes considered to be a fate worse than death. 'Being alive' as such does not count for much.

In view of these kinds of arguments, the American theologian Gerald P. McKenny has characterized contemporary bioethics as a philosophical enterprise in service of what he calls 'the Baconian project'. The Baconian project is the project of modern medicine to enhance control over our bodies as a way of expanding our freedom. Technological control over the body is not just an accidental byproduct of modern medicine, McKenny claims, but must be regarded in view of how our culture understands the human person as a subject.

> We become subjects in part by monitoring, acting on, and exercising vigilance over our bodies, and medicine, especially when it has the role and capacities it has in contemporary society, is a vitally important way in which we carry out these attitudes and performances over our bodies. From this perspective, it is possible to understand how the moral commitments to expanding choice and eliminating suffering depend . . . On the formation of a subject who exercises control over the body.[12]

In the same vein, H. Tristam Engelhardt has claimed that genetic intervention in view of the kind of children people desire to have serves the ideal of controlling natural conditions such that they can become conditions of choice. 'The use of technology in the fashioning of children', Engelhardt writes, 'is integral to the goal of rendering the world congenial to persons.'[13] The world is congenial to persons when the conditions it provides us with are what they would be if we had the power of choosing them. Modern medicine is one of the ways in which our society expands this power. Accordingly, Engelhardt defined the condition of health that medicine seeks to improve as 'the state of freedom from the compulsion of psychological and physiological forces'.[14] Defining health as freedom, Engelhardt seemed to buy into the political notion of life's goodness that has 'choice' as a necessary condition.[15] The point of having a good life, as it emerges from this kind of bioethical theory, is being in control of the conditions shaping my actual life. The point of healthcare provision in liberal society is to enable people to continue to live their lives as they want to as much as possible. Clinical genetics provides just another set of medical interventions that expands the range of possibilities to secure a wider span of control.

Goodness as being

Squarely opposed to this view is the notion that life is good as it is, regardless of whether the conditions shaping it have been subject to choice, either actually or hypothetically. I am particularly interested in exploring this notion as it appears in accounts of people living their lives with a disability, or sharing their lives with a disabled person. In view of their experience, what is there to say for the notion that life is good because of what it is, whatever it may be?

Before we turn to an example of such experience, however, the first task in exploring this question is to distinguish two distinct meanings of the notion that life is good is it is. One way to understand this notion is to say that life as it is is good because being itself is good. This view has been defended in the Thomist tradition, as is well known, on grounds of the claim that goodness and being are really the same.[16] As will become apparent, it is helpful to see how Aquinas defended this notion because it will bring out a meaning that is essentially different from the one I wish to explore. So, the first attempt of clarifying what it means to say that life is good as it is proceeds by way of eliminating what it does not mean.

By way of introducing Aquinas's view on the matter, let me quote the account of it that appears in Etienne Gilson's *The Philosophy of Thomism*:

> Why, Thomas asks, can we say with Augustine that we are good insofar as we exist? Because the good and being are really the same. To be good is to be desirable. As Aristotle says in the *Nicomachean Ethics* (Book I, 1): the good is 'what everyone desires'. Now, everything is desirable insofar as it is perfect, and everything is perfect as it is in act or 'actual'. To be, therefore, is to be perfect, and consequently to be good.[17]

Whatever is is good, because is has actuality, and actuality has (some kind of) perfection, which makes it better than nonbeing, which has no perfection whatsoever. With regard to human being, Augustine has stated that 'inasmuch as we exist, we are good', which Aquinas follows.[18] Human being is good because it is.

There is a question to be considered, however, regarding to what Aquinas means when he says that goodness and being are 'really' the same. The obvious objection to this claim is, of course, that whereas goodness can be more or less, being is always either or.[19] 'Goodness' is a relative term, in other words, which 'being' is not. Some human bodies are healthier and, therefore, function better than other human bodies. Because Aquinas takes 'being' to refer to actual existing objects, it cannot be a relative term in the same sense. An object either exists or it does not exist.

To meet this objection, Aquinas has to allow a difference. 'Goodness and being are really the same, and differ only in idea.'[20] They have the same referent, this is to say, but this does not rule out that the same object can exist in different modes of being. There is a kind of being that is of the either/or kind. This Aquinas calls 'substantial being', or *being simply*.[21] Something has substantial being, or simply is, when it is actual, or exists. At the same time, however, a thing can be a more or less actualized specimen of its kind. A healthy body of a fully grown human being is more developed – 'actualized' – than the body of a toddler, which is why toddlers cannot perform actions that fully grown human beings can perform. This implies a different kind of actuality. Aquinas refers to it as *relative being*. He says: 'By its substantial being, everything is said to have being simply; but by any further actuality it is said to have being relatively.'[22] The reverse distinction obtains with regard to the notion of goodness.

> Goodness signifies perfection which is desirable; and consequently of ultimate perfection. Hence that which has ultimate perfection is said to be simply good; but that which has not the ultimate perfection it ought to have (although, in so far as it is at all actual, it has some perfection), is not said to be perfect simply nor good simply, but only relatively. In this way, therefore, viewed in its primal (i.e. substantial) being a thing is said to be simply and to be good relatively (i.e. in so far as it has being), but viewed in its complete actuality, a thing is said to be relatively, and to be good simply.[23]

So, what we have is that a thing's relative being is a matter of degree, which corresponds with a kind of goodness – 'relative goodness' – depending on the degree of actuality a thing has. Consequently, goodness is predicated as a matter of degree according to a thing's developed state of actuality, which is a mode of being that goes beyond a thing's mere substantial being. With regard to human being, this means that a 'further actuality' refers to a developmental stage wherein a human being is more capable of exercising the capacities that define its nature and, therefore, has reached a more perfect state of being. For Aquinas, of course, the defining characteristics of the human species are reason and will. Because a thing is good in proportion to its capability of exercising its species-defining powers, it follows that the proper exercise of the capacities of reason and will result in *human* goodness.

The conclusion from this analysis can be stated in terms borrowed again from Gilson. 'Just as everything is perfect in the degree to which it is, so also everything is imperfect in the degree to which, in a certain respect, it is not.'[24] Human being is good to the extent that it has actualized its natural capacities; it is imperfect to the extent that it has not. It is not difficult to see how this analysis results in a teleological understanding of the notion of being as goodness. 'Goodness presents the aspect of desirableness', Thomas says, 'Which being does not present.'[25] Desirableness with regard to human being is guided by the proper use of the characteristically human faculties, which are reason and will. To the extent these faculties do not function properly, human being is imperfect.

In view of the present context, this conclusion suggests a problem with the notion of life's goodness as 'being'. Surely human being in the sense of substantial being presents an actualization, and, by implication, a degree of perfection. To be simply human is good to the extent that the act of being – the *actus primus* – is good. At the same time, however, there is goodness proper to human nature consisting in the perfection of its natural capacities, which leads to the judgement of imperfection to the extent that the proper use of these capacities is not actualized. What follows is that Aquinas's account of being as goodness cannot without serious difficulty be true of profoundly disabled human being. The reason is that in the sense of *simply being* profoundly disabled human being lacks the goodness of having actualized its human potential because this potentiality does not exist, as far as we can tell, or exists only marginally. Consequently, in their case, *simply being* implies relative goodness, and even this only in a marginal sense. The reason is that their potential for developing the capacities inherent to human nature is very limited indeed.[26]

The gift of being

Let me turn, therefore, to another meaning of the notion that life is good as it is. If one had to give a single characteristic of this alternative view, one would have to say that it is not framed within the teleology of being as developing human capacities. Instead, it operates within a very different framework, which is that of divine providence. Life is good as it is means that its goodness consists in its being given. In this connection, 'being given' does not merely refer to the factualness of our lives; they are not merely given in the sense in which *data* are given. Instead 'being given' refers to the given as gift in the sense of *donum*. Consequently, the moral point of this alternative view appears as response.

In exploring this view, it occurs to me that we have reached the end of where philosophical reasoning can take us. The reason is that, unless it is used in a metaphorical sense, the language of gift and response implies a giver to whom we respond. The goodness of life as a gift leads to the task of responding faithfully to what is given as a gift. *Donum* instead of *datum*. The crucial difference between these notions is that while 'the given' refers to the anonymous self-generating manifestation of 'being', the 'gift' refers to the personal presence of a being that gives. In short: 'gift' refers to 'giver'. Since the notion of a giver is incompatible with the anonymous manifestation of being, the giver must be identifiable. At this point the chain of philosophical reasoning necessarily arrives at its final stop. The goodness of life as being cannot depend merely on the notion of being a gift, nor can it depend merely on a relation with a giver. Gifts can be bad and therefore no gift at all, so everything depends on what the giver intended to be given. For a gift of life to be really a gift the giver must be identified as good in order that life can be good as it is given. Consequently, the goodness of the gift of life depends on the goodness of the giver.

Speaking non-metaphorically, therefore, the language of life as a gift introduces a religious perspective. Of course, to account for 'life as a gift' one could argue that the gift can be attributed – non-metaphorically – to previous generations that have handed the world to us. What would be lacking in this account, however, is intentionality. From a religious point of view, life as a gift is received from God who intends it to be good. Life's goodness, from this point of view, does not in any way depend on 'having a life' in the sense as explained before because it does not in any sense depend on choice. As a matter of fact, it does not depend on any intrinsic quality at all, because it is grounded extrinsically in the act of God's giving.

From the perspective of our contemporary moral culture this notion must appear as highly contestable, if not downright offensive, and this not only because it completely ignores 'choice' as a necessary condition of the good. Not only does it assume that 'meaning' and 'value' are not derived from the fact that we can choose our own ends, it also assumes that the life's goodness does not reside in particular conditions, such as health, or wealth, or honour, or happiness. Instead it resides in the sole fact that whatever it is that happens in our lives comes from a loving God. There remains the question: is there some plausibility as to the truth of the notion that life is good because of what it is?

A 'presence of peace'

In order to bring about some considerations relevant to this question, let us look at a first-hand account of sharing one's life with a profoundly disabled human being, as was mentioned before. In his book *The Power of the Powerless*, Christopher de Vinck tells us the story of his brother Oliver, a profoundly disabled boy. 'Oliver could do absolutely nothing except breathe, sleep, and eat. Yet he was responsible for action, love, courage, insight.'[27] De Vinck's book is grown out of his memories about Oliver, which he decided to write down only years after his brother had died. He remembers the experience of a certain mystery about the house: 'the house of Oliver', he calls it. It was the mystery of a peacefulness that he attributes to his brother's presence. 'I cannot explain Oliver's influence except to say that the powerless in our world do hold great power.'[28] Oliver could not hold up his head, he could neither crawl, nor walk, nor sing; he never left his bed, he could not hold anything in his hand, nor could he speak. He never knew what his condition was. Yet his brother remembers how they experienced Oliver: 'We were blessed with his presence, a true presence of peace'.[29] Sharing his disabled brother with his family shaped both his life, and his view on life, De Vinck confesses. Part of what Oliver's life taught him had to do with purpose. We set our goals, and make our plans to attain them, but there is a risk in all this activism, De Vinck wants his readers to understand, because it tends to neglect a special kind of attentiveness. We might fail to see 'the hidden inside' of things: the extraordinary that can be found in the ordinary, the profound that is hidden in the trivial.

> I have come to believe we are here to tend to the lilies in the field. We do the best we can. If you have a boy or girl like Oliver in your home, you will know what is best for him or her, for your family . . .
>
> I asked my father, 'How did you care for Oliver for thirty-two years?' 'It was not thirty-two years', he said. 'I just asked myself, "Can I feed Oliver today?" and the answer was always, "Yes, I can".' We lived with Oliver moment by moment.[30]

In his career as a teacher, De Vinck informs us, he often used to tell his students about his brother. He remembers that one day when he was describing his brother's lack of response, he was interrupted by one of his students who raised his hand and said: 'I see what you mean, Sir; you mean he was a vegetable'. 'I presume you could call him a vegetable', De Vinck responded to this student, 'but we just called him Oliver.'[31] De Vinck uses that phrase – 'You mean he was a vegetable' – a number of times in his book as an ominous reminder of how experience makes a difference. However, it is not only experience that matters, because to be able to experience Oliver as 'the presence of peace' you must understand a few things in order to make that experience intelligible. What is it that makes some people see human beings like Oliver as a blessing, where others only see a vegetable?

For someone like Christopher de Vinck the question of whether his brother Oliver has lived a human life is most likely redundant. There is no real difference. Oliver was just Oliver. If there was a question at all, it has been answered by living the life with his family. That life had been lived, as De Vinck's father testifies, in the manner of attending to 'the lilies in the field'.[32] It was a life not lived in the attempt to be in control, but in response to the promise that God will provide. The alternative of seeing Oliver as a blessing, rather than a vegetable, turns on the question of what we take the lives of human beings like him to represent. In the case of De Vinck's family, they regarded Oliver as a child, and a brother entrusted to them to be loved and cared for.

As De Vinck remembers, however, his mother once explained to him that they were blessed by Oliver in ways that were not clear for her at first. This is what she said:

> For many, many years, I was confined to the house, alone and without the support of relatives or friends. My husband was at work all day and

I was with Oliver and the other five children. This enforced seclusion was difficult for me; I had a restless, seeking spirit. Through Oliver I was held still. I was forced to embrace a silence and solitude where I could 'prepare the way of the Lord'. Sorrow opened my heart, and I 'died'. I underwent this 'death' unaware that it was a trial by fire from which I would rise renewed – more powerfully, more consciously alive.[33]

Had Mrs De Vinck pursued the life that she wanted for herself, she probably would not have chosen the life that eventually became hers. As a matter of fact, the woman whose 'restless, seeking spirit' longed for a different life had to 'die' in order for a new self to arise. Mrs De Vinck learned not to resent her task of caring for Oliver as her 'fate'. Clinging to her plans and expectations for herself she probably would not have allowed this to happen. The fact that this role was imposed upon her involuntarily – a fact that has been a major cause of the rise of emancipatory feminism in recent decades – somehow lost its significance. The account of her life with Oliver apparently transcends the absence of choice. She learned to see her life with Oliver as a response, rather than an ill-fated imposition cancelling any prospect for her own plans. Because she managed to learn this, her life with Oliver turned out to be a self-transforming gift, which taught her that accepting the goodness of being as it is created yields an unexpected reward of its own.

De Vinck's account of sharing his family's life with Oliver brings out quite convincingly what it means to respond to the gift of life, whatever it may bring. People who believe in 'choice' as a necessary condition of life's goodness may also recognize, of course, that their life is a gift. If they do, however, they tend to understand this in terms of life's contingencies. 'Choice' is crucial if one believes that the meaning of life is to shape one's conception of the future and turn it into the present. Life's conditions have no other meaning attached to them, following this belief, than to be changed into conditions of choice.

Mrs De Vinck's account is different precisely in this respect. There was meaning to be found in certain things even when she would never have chosen them to be part of her life. In this respect, her experience sheds further light on the distinction between life as given and life as gift. When 'choice' dominates one's moral views, life appears as *datum*, but not necessarily as *donum*. *Datum* refers to 'gift' as contingency; there is givenness, which does not entail there is a giver. In contrast, *donum* refers to givenness that implies purpose. The gift of life implies a giver who may have had a purpose in mind by 'sending' whatever it is that happens to you. This is clearly how Mrs De Vinck came to see her life when she learned to see caring for Oliver as a way of preparing to receive the gift.[34]

In view of these reflections it can be argued, of course, that Mrs De Vinck's account of her own life also can be read in terms of 'choice'. She chose to abandon whatever she may have wanted from her life and decided to accept Oliver. If so read, there is nonetheless an important difference. The difference is that what she expresses and affirms is not a life that has the choosing self as its object. The choosing self is precisely what in her view had to die in order for the new self to rise like a phoenix from the ashes of her old self. In contrast, the conception of life's goodness that makes it dependent upon 'choice' has the choosing self as controlling agent very much at centre stage. This explains why 'choice' in itself is without boundaries. There is nothing, no single condition of human existence that does not present itself to that conception as the suitable object of choice and control. Consequently, this conception of life's goodness implies that 'goodness' is conferred upon our human lives by virtue of our own authorization. This is ultimately what is meant by the claim that people need to be respected as 'the authors' of their own lives. Having a life properly called 'good' they must be in control of how they choose to live it.[35]

Providence

If the foregoing reflections have clarified anything, it must be that a religious notion of life's goodness as being will invoke sooner or later the notion of providence. Since this notion turns on life as a gift – as distinct from the givenness of life – and since life as a gift does not refer to any intrinsic quality of goodness, everything will depend on the nature of the giver. That is to say, everything will depend on who God is and what he does. When it comes to regard the things happening in our lives in this light, theology introduces the notion of divine providence.

The language of providence is not new to discussions on genetics and disability, however, because it has been around for quite a while. Take for example the expressions of being victimized by a genetic disorder as a stroke of 'bad luck', or 'ill fate'. These expressions, as well as talk about the 'genetic lottery', presuppose some kind of view within which the universe is governed by fortune. Ordinarily nature follows reliable patterns, but occasionally there is a disorder that is distributed at random. Some people draw the wrong straw in 'the natural lottery', which, when it occurs, is just a stroke of bad fortune.

The underlying conception of providence guiding this kind of language has its roots in ancient philosophy as an explanation of how individual events are embedded in a larger pattern. How this pattern plays out in the lives of individual human beings is a matter to be decided by the gods, among those in particular the goddess of fortune. In modern times this conception of providence was revived in eighteenth-century deism, which taught that the world was set into motion by God in order to operate according to its own laws. Consequently, as a way of understanding these laws, providence has been replaced by modern science as a much more reliable source of understanding. Since modern science is characterized by the fact that it does not acknowledge purpose in the universe, it is clear that the scientific 'outlook' on the conditions of human life implies that these conditions have no meaning other than the meaning conferred by human purpose. This indicates why the modern scientific perspective is entirely compatible with the conception of life's goodness as dependent upon 'choice' in the sense as explained before.

It occurs to me, therefore, that if there is truth to be found in the notion that life is good as it is, it will invoke a different conception of divine providence.

The providence of deism was providence without Christ. The alternative conception will be one that explains how God the Father of Jesus Christ is present in whatever condition we have to face in our lives. That is what the account of Christopher De Vinck and his mother convey about sharing their lives with Oliver as a way of 'preparing the way of the Lord'. In the process of learning how to do this, they found peace rather than ill-fated tragedy. In the same process, they also found themselves formed and transformed. 'For me, to have been brought up in a house where a tragedy was turned into joy', Christopher De Vinck says, 'explains to a great degree why I am the type of husband, father, writer, and teacher I have become.'[36]

The notion of providence that may teach us to see how life can be good as it is, all of this is to say, requires Trinitarian display. From a Christian point of view, our lives do not simply manifest the continuing reproduction of human being, meaningless in itself except for the meaning we confer upon it. Instead, our lives are properly regarded as a response to the God of whom Christians believe that he will not abandon the work he began. The tradition understood this response in terms of a 'special providence' that taught the believer not only *that*, but also *how* God will provide. God does not necessarily send us what we want, but he surely sends us what we need. What we need, more than anything else, from a Christian point of view, is to be reconciled with him and accept his offer to reconcile us with him in Christ our Lord.

Are these just pious words or do they have any serious bearing on the matter that is our concern here, which is how we respond to being confronted by adverse genetic dispositions? The answer, it seems to me, depends on whether you believe what these words say. People who do, have testified in word and deed that they have found gratitude in their hearts instead of anger, and hope instead of despair. Theirs is a calm and quiet response to what befalls them in life guided by the notion that 'if God the Father is with us, who will stand against us'. It is within this perspective that life appears as good simply because of what it is. The reason is that the most fundamental need of the human heart is met, which is to be at peace with God.

For the moment, there are probably more questions raised than answered by this attempt to an alternative account of the good of being human, but I hope to have indicated how ethical reflection on genetics and disability, particularly intellectual disability, may pursue a different agenda. This agenda presupposes that three consecutive moves are made. The first is to switch from issues about moral taxonomy to issues about the good; the second is to steer away from understanding the human good as conditioned by 'choice' and the third is to switch from the understanding 'being' as a neutral fact in itself – a given in the sense of *datum* – to 'being' as a gift received from God the Father, the Son and the Spirit. If we know how to receive the gift there is every reason to rejoice in our lives, whatever they bring. The reverse is also true: what is not received has not been given.[37] The notion that life is good as it is does not turn on the 'quality' of the conditions of our lives, but on whether we learn to respond to God and find peace with him regardless of the conditions we find ourselves in. To know how to receive this gift is the key to life's goodness, Christianly understood.

Notes

1. Hans S. Reinders, *The Future of the Disabled in Liberal Society: An Ethical Analysis* (Notre Dame, IN: The University of Notre Dame Press, 2000), chs 2–6.
2. See, for example, Gregor Wolbring, 'A Disability Rights Approach to Eugenics', in Sheldon Krimsky and Peter Shorett (eds), *Rights and Liberties in the Biotech Age* (Lanham, MD: Rowman & Littlefield, 2005), pp. 146–51; Eric Parens and Adrienne Asch (eds), *Prenatal Testing and Disability Rights* (Washington, DC: Georgetown University Press, 2000); Anita Silvers, David Wasserman and Mary Mahowald, *Disability, Difference, Discrimination: Perspectives on Justice in Bioethics and Public Policy* (Lanham, MD: Rowman & Littlefield, 1998); Susan Wendell, *The Rejected Body: Feminist Philosophical Reflections on Disability* (New York: Routledge, 1996).
3. See, among many others, Paul K. Longmore, *Why I Burned my Book and Other Essays on Disability* (Philadelphia, PA: Temple University Press, 2003).
4. The notorious exception comes from the deaf community. A famous case was described by Andrew Solomon, 'Deaf Is Beautiful', *New York Times Magazine*, 28 August 1994, p. 67. For the notion of 'deaf culture' see Nora Ellen Groce, *Everyone Here Spoke Sign Language: Hereditary Deafness on Martha's Vineyard* (Cambridge: Cambridge University Press, 1985).
5. This can be inferred from the fact that people with intellectual disabilities are hardly ever chosen as friends. Their kind of life is not the kind of life with which people want to be associated. For a reflection on this phenomenon from within disability studies see Anne Louise Chappell, 'Still out in the Cold: People with Learning Difficulties and the Social Model of Disability', in Tom Shakespeare (ed.), *The Disability Reader: Social Science Perspectives* (London: Cassell, 1998), pp. 209–20. For a sociological survey see Phyllis A. Gordon, Jennifer Chiriboga Tantillo, David Feldman and Kristin Perrone, 'Attitudes Regarding Interpersonal Relationships with Persons with Mental Illness and Mental Retardation', *Journal of Rehabilitation*, 70.1 (2004).
6. The most successful example of this approach to ethical reflection remains Tom L. Beauchamp and James F. Childress, *Principles of Biomedical Ethics* (New York: Oxford University Press, 4th edn, 1994).
7. The notorious exception here has been Lee Silver's book, *Remaking Eden: Cloning and beyond in the Brave New World* (New York: Avon Books, 1997).
8. The best way to sustain this claim is to read autobiographical narratives written by people who want to share their own experience. Among my own favourites on intellectual disability are Michael Dorris, *The Broken Chord* (New York: Harper, 1989); Christopher de Vinck, *The Power of the Powerless: A Brother's Legacy of Love* (New York: Doubleday, 1990); Kenzaburo Oë, *A Healing Family* (Tokyo: Kodansha, 1995); Michael

Berube, *Life as We Know It: A Father, a Family, and an Exceptional Child* (New York: Vintage, 1996); Martha Beck, *Expecting Adam: A True Story of Birth, Rebirth, and Everyday Magic* (New York: Berkley Books, 1999); Rachel Simon, *Riding the Bus with my Sister* (Boston, MA: Houghton Mifflin, 2002). Collections with short stories are: Donald J. Meyer, *Uncommon Fathers: Reflections on Raising a Child with a Disability* (Bethesda, NY: Woodbine House, 1995); Stanley D. Klein and Kim Schive, *You Will Dream New Dreams: Inspiring Personal Stories by Parents of Children with Disabilities* (New York: Kensington, 2001).

9. James Rachels, *The End of Life: The Morality of Euthanasia* (New York: Oxford University Press, 1986).

10. This kind of logic is found in many works of bioethics since the 1990s. See, for example, Helga Kuhse and Peter Singer, *Should the Baby Live?* (Oxford: Oxford University Press, 1985); John Harris, *The Value of Life: An Introduction to Medical Ethics* (London: Routledge & Kegan Paul, 1985); Allen Buchanan and Dan Brock, *Deciding for Others: The Ethics of Surrogate Decision-making* (Cambridge: Cambridge University Press, 1989).

11. Harris, *The Value of Life*, pp. 14–19.

12. Gerald P. McKenny, *To Relieve the Human Condition: Bioethics, Technology, and the Body* (Albany, NY: State University of New York Press, 1997), p. 216.

13. H. Tristam Engelhardt, Jr, *The Foundations of Bioethics* (New York: Oxford University Press, 2nd edn, 1996), p. 239.

14. H.T. Engelhardt, 'The Concepts of Health and Disease', in Arthur L. Caplan, *et al.* (eds), *Concepts of Health and Disease: Interdisciplinary Perspectives* (Reading: Addison-Wesley, 1981), pp. 31–45, 43.

15. Ascribing this view to Engelhardt is contestable in the sense that, particularly in his earlier work, Engelhardt claimed to be exploring a conception of health-care and health-care ethics in the context of a society characterized by moral diversity and ethical pluralism. He may therefore want to reply that no conception of the good life that does not have 'free choice' as a necessary condition can be defended by the rules of public justification in a pluralist, secular society, but that this does not mean that he himself subscribes to such notion. In his later work, Engelhardt is much more explicit about the difference between what he thinks can be publicly defended and his own orthodox beliefs about the good of human being. For example, he opposes the view under consideration here as 'liberal cosmopolitanism', which he characterizes in the following way:

> Because of a search for meaning in a fully secular context, a liberal cosmopolitanism is embraced, which assigns lexical priority to liberty and to equality of opportunity. Persons without anything to value beyond themselves come to value first the opportunity to shape their own lives' (H. Tristam Engelhardt, Jr, *The Foundations of Christian Bioethics* [Lisse: Swets & Zeitlinger, 2000], pp. 40–45, 43).

16. Aquinas, *Summa Theologica*, Ia, 5, q1.
17. Etienne Gilson, *Thomism. The Philosophy of Thomas Aquinas* (trans. Laurence K. Shook and Armand Maurer; Toronto: Pontifical Institute of Medieval Studies, 6th edn, 2002), pp. 100–101.
18. Aquinas, *Summa Theologica*, Ia, 5, q1; Gilson, *Thomism*, p. 100. Augustine's statement is found in *De Doctrina Christiana*, Book I, 32.
19. Aquinas, *Summa Theologica*, Ia, 5, q1, obj. 3.
20. Aquinas, *Summa Theologica*, Ia, 5, q1, resp.
21. Aquinas, *Summa Theologica*, Ia, 5, q1, ad1.
22. Aquinas, *Summa Theologica*, Ia, 5, q1, ad1.
23. Gilson, *Thomism*, pp. 100–101.
24. Gilson, *Thomism*, p. 100.
25. Aquinas, *Summa Theologica*, Ia, 5, q1, resp.
26. For an extensive discussion of this point with Roman Catholic moral theology, see Hans S. Reinders, *Receiving the Gift of Friendship: Profound Disability, Theological Anthropology, and Ethics* (Grand Rapids, MI: Eerdmans, forthcoming). In that discussion I develop the claim that the framework of Aristotelian–Thomist metaphysics creates the danger of an anthropological 'minor league' for human beings with profound disabilities.
27. De Vinck, *The Power of the Powerless*, p. 31.
28. De Vinck, *The Power of the Powerless*, p. 32.
29. De Vinck, *The Power of the Powerless*, p. 31.
30. De Vinck, *The Power of the Powerless*, p. 32.
31. De Vinck, *The Power of the Powerless*, p. 31.
32. In this phrase De Vinck is referring to the gospel of Matthew.
33. De Vinck, *The Power of the Powerless*, p. 94.
34. I am referring to the remark she makes about expecting Christ where she says: 'I was forced to embrace a silence and solitude where I could "prepare the way of the Lord"'.
35. An all-important implication of this concept is that it excludes all those incapable of purposive agency from participation in the human good. That is to say, it excludes those human beings who because of their impairment cannot affirm their own being. Consequently, if there is to be an inclusive account of human being, it cannot depend in any way on the centrality of the choosing self. In this respect, the lives of persons with profound intellectual disabilities pose a serious challenge to our ethical thinking. (This challenge provides the subject matter of my book *Receiving the Gift of Friendship*, forthcoming).
36. De Vinck, *The Power of the Powerless*, p. 31. For another account that has providence in parenting a disabled child as one of its major themes, see Beck, *Expecting Adam*.
37. Henry Nouwen, *Adam: God's Beloved* (Maryknoll, NY: Orbis, 1997), p. 31.

11

Angels with Clipped Wings: The Disabled as Key to the Recognition of Personhood

Bernd Wannenwetsch

This chapter suggests a shift of parameters in the debate on 'disability and personhood'. I start by making a case for the abandoning of the 'inclusion' discourse towards an understanding of the 'recognition' of personhood as a moral challenge of a different sort. Instead of asking why and under what circumstances disabled human life could be said to fall under the descriptor 'person' as a concept whose parameters are fixed in advance and in abstraction of their lives, I suggest we better understand the disabled as key to the process of recognizing personhood in general. Drawing on the philosophical anthropology of Helmuth Plessner and Robert Spaemann, I shall show (in section 2) that the decisive point for the understanding of 'person' is in the phenomenon of human kinship and in the peculiar 'positionedness' of human life as being oriented towards a *Mitwelt* of other human beings that claim us by the sheer virtue of existence.

These insights from philosophical anthropology will prepare us well to a more concrete fleshing-out of the dual nature of the act of recognition. With the help of the Biblical notion of 'flesh from my flesh', I shall demonstrate (in section 3) that according to the logic of the Genesis narrative, the revelatory moment in Adam's recognition of 'the other' in the first human–human encounter is at the same time the disclosure of his own humanity to himself that is provided through the other human being once recognized as kin. In a next step (section 4) the argument must, however, take account of the inner resistance that this recognition process meets and the state of oblivion it enters, if the 'objects' cease to be conventional or present a challenge of the sort disabled human life unfailingly does. Precisely in reckoning with such tendencies, we can then portray the lives of the disabled as having a genuinely angelic character and mission (section 5). As our kin they help 'us' – those who consider themselves normal in terms of the deficiencies we find in them – to understand our own human nature in terms of the same creaturely dependency that the disabled demonstrate us in degrees that are impossible to overlook. Impossible to overlook, they still require 'different eyes' (section 6) to really see and understand without the distortion fear brings to human sighting. Finally, I will summarize how all these considerations help us reframe the question of 'personhood and the disabled'.

1 Overcoming 'inclusion'

Responses to the 'disability problem' often revolve around the question of whether or why we should speak of disabled human beings as persons. From the perspective of the Christian tradition it seems imperative to agree with those who, in their desire to position themselves over against thinkers such as Peter Singer[1] and their claims, have argued that we must. Yet, the same desire to position oneself in such a way may make us prone to overlook that a certain kind of moral discourse, to which Christians can easily be tempted, is almost as ill-fated as the positions they want to overcome. A phenomenological analysis of the language act of addressing someone as 'person' will demonstrate that the 'not yet' or 'no longer' which Singer and his like promote with regard to the inclusion of the disabled under this notion is rather mirrored than overcome by the emphasis 'still'[2] in the sort of critical reaction these accounts often provoke.

Since being referred to as persons, as well as referring to others by this notion, is something we usually take for granted, we are tempted to understand the predication of a disabled human being as 'person' as a kind of benevolent stretching of the concept from the usual case toward the unusual. In other words, we are led to think of this 'including' of the disabled within the protected zone which we inhabit – 'even the disabled!' – as a required *moral act*. Christian contributions to the discussion have often attempted to provide reasons for the adequacy or even necessity of this moral act of inclusion. Even though disabled people may lack one or more of the features that we usually expect to perceive in what we call a person, for God's sake, for creation's sake, for the sake of society's moral quality, etc., the claim goes, we need to stretch the notion so as to encompass even the most severe forms of defective human life. If, for example, the concept of person draws on the capacity for intentional behaviour, then we are summoned to demonstrate that the human life under consideration is indeed showing sings of intentionality – however vaguely (in the case of the disabled) or from surprisingly early stages on (in the case of embryos in the womb). Trust must then be put in the growing refinement of our perceptive instruments that allow us to observe or detect the features we think necessary for the inclusion.

In this chapter, I do not aim to assess individual strategies of this type and their respective merits or limitations. I rather wish to call attention to the mistaken assumption on which they are based: the need for an act of inclusion that, in turn, rests on an abstract and preconceived concept of personhood that is to be merely applied to disabled human life, instead of being won from a perceptive understanding of the phenomenon itself.

In contrast to the silent assumptions of the inclusion discourse, I shall argue that the disabled must not be perceived as living at the margins, but at the centre of any sound philosophy of personhood. In other words, they provide not just a test-case but the very paradigm for our recognition of any person. 'Person' is a word that functions as 'moral notion' in terms of a *nomen dignitatis*. Speaking of a 'person', we recognize that we are dealing with a being that is to be treated as *somebody* instead of a *something*. In ascribing the term 'person' to someone, we dignify that individual even as we find ourselves under the particular moral claim that this notion conveys. Yet the idea of 'ascribing' remains easily entrapped in the same one-sidedness as the notion of 'including'. What it overlooks is the fact that we cannot simply mean to bestow the dignifying term 'person' on someone or withhold it from someone without our own dignity as a person being at stake.

My analysis will focus on the simultaneity in which we *discover and prove ourselves as persons* (anew or for the first time in a deeper sense) when recognizing the personhood of the other in addressing him or her in this fashion. The question is not whether or to what degree we are happy to ascribe personhood to others, but rather whether and to what degree we live (and understand) our own lives as those whose dignity as genuine human beings is at stake in the recognition of personhood in others.

If we wish to transcend the ill-fated discourse of 'inclusion' and 'ascription', we must be sensitized to the role that discovery or even revelation plays in this recognition process. In the same way in which a silly, 'fun-loving' teenager might find herself miraculously transformed into a responsible person through the developing human being in her womb or in her arms, the recognition of personhood is principally a matter of dual (though necessarily mutual) rather than of unilateral recognition or even ascription. Disabled human life provides the very paradigm for this, in so far as we learn precisely from those supposedly 'extreme' cases in which a disabled child effectively bring us (back) in contact with our own humanity, that in spite of its 'conventional' surface, the language game of 'personhood' draws its grammar from revelation-like discovery and personal transformation.

To the degree in which the philosophical discourse on this matter has predominantly dwelt on argument rather than phenomenon, on 'whether' and 'why' rather than on 'how', this essential revelatory moment has been neglected to the peril of both discourse and the actual standing of disabled persons in Western societies. Yet there are notable exceptions. As for their particularly helpful insights in the context of our enquiry I wish to draw attention to the philosophical anthropology of two German thinkers: Helmuth Plessner and Robert Spaemann. While Plessner's work explains the way in which personhood is rooted in a unique triploid position in which human beings find themselves as a matter of their very existence, Spaemann's book *Persons* helps clarify that there cannot be any other basis on which we recognize each other as persons than our mutual belonging to the same species.

2 Positionedness and kinship

Helmuth Plessner's way into philosophy was unusual, as his first career was in zoology; however, his lifelong fascination with the quest for the genuine nature of the animal *Homo sapiens* provided the driving force behind his philosophizing.[3] His enquiries into the 'stages of life', as one of his major studies was titled, were driven by a conceptual curiosity in regards to the way in which the different forms of organic life are organized.

In contrast to the 'open' form of organization in the world of plants, and the 'centric' existence of the beasts, Plessner found that human beings live 'excentrically' – eccentricity being just another name for personhood. A 'person', as he defines the concept, is a bodily being that is characterized by a threefold positionedness: *as* body, *inside* a body (traditionally, 'soul'), and *relating to* its body as if from the outside (traditionally, 'spirit').[4] The 'spiritual' capacity of a human person to relate to him- or herself is represented as a peculiar 'space' that Plessner calls *Mitwelt*. Rather than simply surrounding the human life *as Umwelt* (environment), *Mitwelt* is a qualified sphere that claims the human person socially and morally. It is not established or created by human interaction, but must be understood as part-and-parcel of the unique positionedness of the human existence itself: '*Mitwelt* is the sphere of other human beings that man perceives as the form of his own position'.[5] In this capacity, the *Mitwelt* can be said to 'carry' the person as much as it is carried by the person. 'In between I and I, I and thou, I and him or her, there lies the sphere of the spirit.'[6]

One of the advantages of Plessner's anthropology is that it differentiates between consciousness, soul and spirit. It does so in such way as to clarify the two latter constitutive dimensions as givens which come with the peculiar position that characterizes human life prior to any activity and independent of any developmental stage of the individual.

> The human life exists as internal world [soul] independently of whether he knows of it or not. In the reflection of the mind it is only made present to him . . . In order for the self to be able to relate to himself as a reality *sui generis*, it must presuppose his very being as extrinsically organized.[7]

And in the same fashion, Plessner puts it for the spirit: 'Man possesses spirit in the same way as he has a body and a soul. He has them, as he is them and lives them. Spirit is the sphere by virtue of which we live as persons, in which we stand, precisely because our positionedness entails it.'[8]

Although Plessner's sophisticated differentiations would require and deserve a more detailed analysis than we can give here, it should be evident where the relevance of his anthropology lies for our questions on personhood and disability. Whereas most theories of personhood render constitutive the capacity to relate to oneself, the problem is they often do so in a shortcircuited fashion, on Plessner's view embracing but one of the positions he discusses. Most commonly they draw on reflection as an instance of consciousness and overlook the two other ways in which the human being is positioned in regards to him- or herself as body and inside the body (soul). As we understand this onesidedness, we are led to understand another hypertrophy it engenders, in that the whole emphasis in the idea of personhood must now fall on the capacity of active *relating* (establishing relationships through intentional activity), whereas any form of *relatedness* as a result of the unique position that human life assumes slips out of the horizon of the concept of personhood. As Plessner's considerations on *Mitwelt* suggest, primal relatedness is by no means a passive concept but very much involves relating on a sliding scale of activity. But it makes all the difference whether we understand our capacity to relate as establishing a world of relations or, alternatively as Plessner suggests, as an inherent feature of the trilateral position that characterizes human life prior to action and reflection.

For Plessner, who was already in his nineties when he compiled the book from which we quoted, 'personhood' was not a troubling concept in its own right. It was basically a name to describe human nature when understood in its fullness and uniqueness that set it apart from other animals and organic forms of life. Yet, in its consistent attempt to be faithful to the phenomena of human existence and the refusal to give in prematurely to the attraction of wornout conceptual dichotomies such as naturalism vs realism, Plessner's philosophical anthropology prepares the ground for our turning to Robert Spaemann's work. Although Spaemann's approach takes the opposite direction, in eliciting the inner logic of the concept of 'person' as it actually functions in human communication, it can be seen as complementary to Plessner's work in a variety of ways.

In his seminal work *Persons*[9] Spaemann gives an compelling account of the history, role and significance of the concept of personhood, which is refreshingly sensitive to the dialogical nature of the notion and the revelatory dimension in the recognition of this status. Particularly helpful in the context of our enquiry is Spaemann's clarification of the non-generic nature of the concept of person by which he undermines the nominalist logic that grounds the idea of 'speciesism' which Peter Singer so wants to reject. Singer's distinction between persons and members of the human species mistakenly assumes, as Spaemann indicates, that 'person' denotes an instantiation of an abstract concept to which individuals may be said to belong by virtue of the possession of certain qualities claimed as constitutive of the concept. In contrast to this, Spaemann maintains: 'If being a person is a *modus existendi*, there is no *genus proximum*, no more general category that the concept of "person" could specify'.[10] Rather than being an instantiation of a concept or a member of a class, a person is 'but a participant in a community of mutual recognition'.[11] There are no 'potential persons' any more than there can be only the 'idea' of a person. 'Person' as a concept exists only as long as there are individual persons.[12] If 'person' is not a generic term but rather the way in which individuals of the human genus exist, Spaemann goes on, then 'each of these individuals occupies an irreplaceable position in the community of persons, which we call mankind'[13] – and does so not by co-option but by birthright. 'There can, and must, be one criterion for personality, and one only; and that is biological membership of the human race.'[14]

This claim of belonging to the human community of persons from birth is not subject to the charge of 'speciecism', since this belonging is not – as in the world of inanimate objects and artefacts – established on the basis of likeness, but through genealogical relations of 'kinship'. All human beings are related to one another as descendants of a single woman who, as current scientific theories suggest, lived about 200,000 years ago.

> Members of the species *homo sapiens* are not merely exemplars of a kind; they are kindred, who stand from their outset in a personal relation to one another. 'Humanity', unlike 'animality', is more than an abstract concept that identifies a category; it is the name of a concrete community of persons to which one belongs not on the basis of certain precise properties objectively verified, but by a genealogical connexion with the 'human family'.[15]

These genealogical relations are not merely 'biological facts' but establish cultural and social relationships that carry specific moral expectations with them. Characterizing the dialogical quality of 'personhood' as a matter of mutual recognition, Spaemann (building on Boethius's classical definition of person as 'individual substance of a rational nature') defines 'person' as an individual member of the human race that 'has' his or her nature rather than simply 'being' it. To be a person means to relate to one's own nature in an intentional and hence morally qualifiable way.[16]

In the light of these clarifications, Spaemann speaks of the existence of the severely disabled as the 'acid test of humanity'.[17] Due to their lack of expressive capacities, we perceive their nature as 'broken', as we perceive a broken chair still as a chair.

> A disabled person may lack such properties. And it is those who lack them who constitute the paradigm for a human community of recognising selves, rather than simply valuing useful or attractive properties. They evoke the best in human beings; they evoke the pure ground of human self-respect. What they give to humanity in this way by their demands on it is more than what they get.[18]

Spaemann's intriguing formulation in the last paragraph provokes further questions: are the disabled really evoking the best in us? And if so, how does it happen, and what will this 'best' consist in? The first question invites an objection: is this not a rather optimistic statement? At the very least, we will have to strive for an account of the recognition of personhood that is capable of explaining why this is sometimes, but by no means always or automatically, the case. What is it in the process and act of recognition that can go so horribly wrong and evoke the worst rather then the best in us?

And for the happy case in which the 'best in human beings' is actually brought out: Who, then, is to be credited? Is it the power of the disabled person herself, the power of the encounter, or the actualizing of the agent's own moral power that is merely occasioned by the encounter with a needy human being? And how are we to spell out the idea of 'evoking'? What does it mean to 'evoke the pure ground of human self-respect'? What, in other words, is the 'business' or – traditionally speaking – the 'mission' or 'calling' of the disabled within the wider family of humans in which the majority of people understand themselves as lacking the lack that they find in the disabled?

A first, immediate, answer to this latter question would of course point out, as Spaemann himself does,[19] that what disabled persons have to give to their fellow human beings who lack certain things is precisely the provision of occasion for help and caring that their neediness evokes. This thought is in line with a well-established argument in the theological tradition according to which the *raison d'etre* and the mission of the poor has always been to enable the affluent to do good works by turning to their needs. Significantly, though, the rationale of the enabling of good works and the easing of the lives of the poor was never a stand-alone reason. It was typically accompanied by considerations relating to the life of the giver. In this respect, the mission of the poor was, so to speak, to warm the heart of the affluent and thereby assist God in saving them from self-centredness and idolatry.[20] In this vein it was made unambiguously clear that the relation between the privileged and underprivileged, theologically considered, could not be unidirectional and condescending, but needed to be understood in terms of mutual bestowal and blessing.

With this theological standard of a spiritually mediated mutuality in mind, we can now attempt to find a more nuanced answer to the question of the specific mission of the disabled within the human family. For this purpose, we first turn to the archetypal biblical account of the recognition of personhood, as we find it in the (second) creation narrative.

3 'Bone from my bone'

'This at last is bone of my bones and flesh of my flesh' (Gen. 2.23). The man's cheer of excitement and delight when being introduced to the woman is more than simply the recognition of the other sex. It is the self-recognition of the man who recognizes himself *in* the other sex. More precisely, through the woman, man is provided with a revelatory disclosure of his humanity, that is, of his human nature as a reasonable and social animal.

Such recognition, as the narrative suggests, could not be provided through the animals, but only by bone and flesh of another kind: 'Finally, bone from *my* bone and flesh from *my* flesh'. Although man (*ha adam*), when faced with the animal world, was able to recognize his difference from them, they could not provide him with the positive recognition of his own nature. They shared only aspects rather than the full manifestation of his nature, and they are subjected to mankind in his dominion over them. One expression of this rule over the animal world is man's power to 'give names' to the animals (Gen. 2.19–20), which includes determining the respective level of proximity or non-proximity to himself according to utility or other calculations. Within the family of mankind, a name can be given to the newly found partner only according to the *actual status* of the shared nature (*adam – adamah*). In contrast to the dominion over the animal world, no such hierarchy and therefore defining power is given.[21] It is not up to man to define humanity in this way,[22] whether in the general or in the particularity of any individual member of the human race. Among humans, the articulation of belonging as expressed in the language of personhood cannot be accorded or even granted, only recognized. And this recognition is a process of critical interdependence, in which we discover ourselves not only simply *through* the other (dialogical: self), but actually *in* the other (kinship: nature).

It is no incident that this biblical insight, although of validity for all human relationships is, in the first instance, mediated through the relationship of the sexes whose difference is manifested in their *bodies*. It is precisely as bodily beings that humans experience their mutual dependency, both in the most pleasurable and most painful way. The recognition of human dignity, whether focusing on the notion of 'person' or on the concept of the 'image of God', must be mediated through, and not abstracted from, the fact the human beings are bodily beings.

As Dietrich Bonhoeffer put it in his *Creation and Fall*:

> humankind created in this way is humankind as the image of God. It is the image of God not in spite of but precisely in its bodily nature. For in their bodily nature human beings are related to the earth and to other bodies; they are there for others and are dependent upon others. In their bodily existence human beings find their brothers and sisters and find the earth. As such creatures human beings of earth and spirit are 'like' God, their Creator.[23]

In the light of this biblical insight, we can now see that the understanding of 'person', as advocated by Singer and his allies, is one that draws its rationale precisely from the *neglect of the cultural and moral meaning of the generative web that marks the human species as kinship*. Given this fact, it seemed like a matter of divine irony that more recently provoked Peter Singer to advancing practically 'speciesism', when he, contrary to his voiced opinion that every person is to be treated to the exact same standards of objective needs and desert, was reported to have put a large sum of money into special care provisions for his elderly mother who was suffering from Alzheimer's disease. What the philosopher recognized in the *special* case of his own mother,[24] whom he felt he must treat as a person irrespective of her declining ability to demonstrate the traits which her son's definition associates with this notion, is actually the key to the matter *in general*. We are, in fact, to treat every member of our species as a person, precisely because they are 'family', because we share with them the same human nature, the nature of which they, in turn, help us understand in a fuller and better way.

4 Oblivion

I had my abortion on June 30th, and I was a mess. I was weeping all the time . . . We were watching the parade on Main Street in Hamlet, at my in-laws' cottage, and a family with a kid with Down's was standing in front of me. Right there at the parade, honest to God, like a sign direct to me. And the thing was, I really looked at the kid, how she dripped her ice cream all over, how she couldn't be made to do what the other kids wanted. I looked at her and thought, 'she doesn't belong in that family'. She didn't look like them, she looked like someone else. Like a lot of someone elses, not quite from the same race, if you know what I mean. And it made me feel, well, that I'd done the right thing, that the one I aborted wasn't quite from my family, either.[25]

It marks the human condition under the Fall that the recognition of kinship and its concomitant moral claims are no longer a matter of course, but seem at times exceedingly difficult. On the surface, the complicating factor apparently lies in the sheer size of the human family in which the sort of bonding that a small family unit 'naturally' evokes and enables, cannot not be taken for granted in anything approximating this degree.

As Hannah Arendt[26] noted in respect to the fate of 'displaced' persons in the wake of the Second World War, the sheer fact that humans share the same nature by no means prevents them from treating other members of their species with the utmost disrespect and cruelty. This sort of 'implicit knowledge', to quote Michael Polanyi,[27] usually suffices to govern our day-to-day conduct in a way that is compatible with the respect due to other human beings, but it often ceases to do so once we see our own life, health, status or well-being threatened. Anxiety and greed, amongst other things, have the notorious capacity to cast the spell of oblivion on members of the human race to make them 'forget' their common belonging.

This is why, in a state of crisis and oblivion of shared humanity, this implicit knowledge needs to be buffered by patterns of explicit and institutionalized reminders, such as the Geneva Convention: 'Look, this prisoner of war, this terrorist, is still a human being and deserves to be treated as such, and not have a lead put around his neck, be stripped naked, sexually humiliated, and so on'.

Such legal and conventional measures function in a similar 'demonstrative' and instructional way as God's proactive presentation of the woman to the man. Though *ha adam*'s cry 'bone from my bone' sounds like an utterance of immediate and unambiguous recognition, we must not forget that, according to the biblical narrative, it took God's preceding initiative of presenting the woman to the man in the first place. Revelation never occurs just 'in the eye of the beholder' or exclusively in the object of divine presentation itself, but always in between as a matter of divine interference. This is why recognition of personhood is, ultimately, a spiritual rather than an empirical matter.

If the recognition of personhood is not just parallel to but essentially one with the recognition of 'family' in the morally qualified sense of the notion, it would seem obvious that the paradigmatic experience of kinship as it is inscribed into the reproductive pattern of so called 'core' family units that actually live together, grounds much of our ability to recognize personhood.[28] It does so not least by creating and recreating the implicit knowledge on which we usually rely. To live in a family and learn to recognize those who belong to it is an elemental and elementary schooling in the recognition of personhood.

A three-year-old whose newborn sibling arrives in the family home for the first time may find it very hard to relate positively to the odd-looking creature in the cradle who ends his unrivalled status as the centre of attention. 'It' does not at all look like him or like mum or dad, nor does it behave in a way that would recommend it to be counted 'one of us'. At least initially, it is not for any perceivable quality or characteristic of the newborn that the older sibling is expected to respect the invader and to learn to treat him as family, but for the sole reason of being told to: 'Look, this is your brother Jack'. This is the message that he will pick up both implicitly and explicitly, by way of observing the parents' affectionate ways with the newborn that resemble the way they treat him, and by being patiently taught by them in a variety of circumstances – upon his first encounter with the newly arrived and whenever his attitude towards him necessitates a reminder.

Any sound and realistic account of 'personhood' needs to come to terms with the complication that although kinship is the 'natural fact' that begs to be acknowledged, the language-game of 'personhood' does not come naturally to us. It is not 'natural' even within the context of the core family unit, but has to be learned by instruction, through exposure to a kind of *verbum externum*. To speak of a learning process in the recognition of personhood means to be aware of the various strategies of resistance and the tendency towards oblivion that make this process both necessary and difficult. Such tendencies, and hence the need to learn, are by no means restricted to intra-familial scenarios, but apply to larger social units as well as to societies at large.

The revelation of kinship and belonging to 'one family' is never unilateral as uncovering of the other as kin alone but always dialectical: we discover our own humanness in the humanness of the other. This is why the process of recognition is both natural yet does not come naturally to us. If it is true that we are faced with what is at stake in our own dignity as human persons when we look at the other whose 'personhood' is at stake, it seems just too natural to conceive of the 'criteria' to apply to the other in such as way as to satisfy theimage we harbour of ourselves. This is particularly obvious in the case of the disabled. What they reveal to us when discovered as our kin is precisely that truth about our own human nature which we tend to ignore, deny and ward off: that we are, after all, not simply rational animals, but 'dependent rational animals'.[29] It is precisely to the degree, in which the lives of these sons and daughters, brothers and sisters, aunts and uncles, emphasize human dependency and need, that they unfailingly shock and perplex us, since we have invested a whole world, both individually and culturally, to cover our existential nakedness with the cloth of achievement, power and control.[30]

5 Angelic mission

The disabled do not automatically evoke the best in us, but their recognition as persons functions as a litmus test for our own human dignity as persons. They function as such because their existence as fellow human beings and our kin can only be genuinely recognized if, in turn, we recognize the revelatory quality of their lives for the understanding of our own humanity. It is precisely what they seem to lack that reveals a truth about our own life which, in turn, possesses the potential to heal us from the deceptive love-affair with the epochal demons of self-mastery and control. The disabled and retarded are to be counted one of us not *in spite* of their 'embarrassing' features and deficiencies: this would render recognition a condescending moral act of stretching the defining boundaries of our own self-image as self-reliant masters of our own lives.[31] Not in spite of her embarrassment are we to count a disabled person as one of us but precisely *because of* her embarrassment – an embarrassment that needs to be recognized as our own, as the hidden and notoriously unacknowledged fact of our own lives as *enoschim*, creatures of the dust.

If the disabled convey this existentially crucial message to us and reveal to us the truth about our own life, we need not shy away from recognizing a religious concept here: they actually act as *angeloi* – God's messengers that sometimes come in strange, yet familiar cloth, both from within and outside our human world. By characterizing the disabled as potential angels, I do not wish to distance them yet again from the realm of humanness or to bestow a romantic quality on their lives.[32] As I hope to have made unambiguously clear by now, if they are considered angelic it is precisely for their human nature and the revelatory dimension of our sharing in it as their kin. The disabled must not be romanticized as innocent or even 'better human beings'[33] but need to be understood as humans with particular qualities and a particular mission. Our disavowal of this romantic account aims precisely at *not* denying the sense of alienation that disabled human life usually provokes in those who consider themselves 'normal'.

Angels are powerful yet fragile beings. This is why they are often depicted with wings. The message that they bring can, notwithstanding its divine authorization, be resisted. Angels can be mistaken and attempts made to abuse them (Gen. 19.4f). If we understand the particular mission of the disabled in the angelic way described above, it must be added that the message shares the fragility of the messengers. Our kinship with the disabled gestures towards our own innate and insurmountable state of dependency as 'children' (*ben adam*) of God, but there is nothing self-evident in this message that would assure it to be listened to, believed and incorporated in the life of a society.

In our times in particular, it would seem rather apt to say that these angels live among us with clipped wings. Yet what clips their wings and 'disables' them in this theological sense is our society's obsession with the images of mastery and autonomy. The wings of these angels are clipped to the degree in which we, ideologically or practically, bring us in a safe distance from them, outside of the sonic radius of their voices, so to speak. These angels cannot fly after us;

their mission for us depends on our willingness actually to stay within the realm of encounter as to have them 'in our midst'. When it comes to understanding their lives and our own lives as mirrored and qualified by our kinship with them, there is no surrogate to the actual sharing of our lives. We need concepts such as 'personhood' to help us define what human life is about, but if we do not open up our conceptual thinking to the challenge that the phenomena (the living *among us* of those whose dependency cannot be overlooked) provide, we will end up worshipping the straitjackets we wear, both intellectually and practically.

It is because of the potentially alienating quality of the encounter that angels usually begin to address humans with the words 'Fear not!' There is much in the life of the disabled, for some even its very essence, that causes fear in those who encounter it. That fear is often rationalized in terms of the uncertainty to 'cope' with the strange and unfamiliar in the disabled person. But at a deeper level it mirrors the anxiety in regards to what the disabled reveal about our own life in terms of what we share with them in one and the same human nature.

6 With different eyes

As Helmuth Plessner demonstrated so convincingly even for the sphere of sensual perception, there needs to be distance not only to the perceived object but also to our familiar modes of perception: 'We only perceive the unfamiliar.' In a most helpful way, Plessner explains the epistemological dialectic between *Anschauung* (perception) and *Begriff* (concept) for the structure of human experience: Only the unfamiliar and even alienated can be perceived (*zur Anschauung kommen*), while only the familiar can be conceptualized.[34]

As for the concept of personhood, we may conclude that its development needs to be fed by a mode of perception that is open to the perplexing impact of the unfamiliar within the familiar. 'Pain is the eye of the spirit',[35] Plessner writes. In order to understand our own life and world, we depend on the exposure to the unfamiliar, unpleasant, shocking dimensions of our reality. 'The shock-condition-ness of the perception that we have in mind here, the reliance on shock and pain of the alienated gaze is one of the a priori conditions of understanding.'[36] It is this alienated gaze that makes the familiar unfamiliar which eventually enables conceptual thinking to redescribe the familiar in a more truthful way. We need to be equipped with 'different eyes' (as Plessner's book is aptly titled) if we wish to see our human life for what it is rather than what our fear-driven first impulses want it to be.

This is the angelic mission of the disabled we have in mind. Through the alienation that we feel at the sight of them, they make us see ourselves in a different light, as our gazing at them is reflected back to us. In this way, they bestow on us the 'different eyes' that help us perceive our humanness as 'dependent animals'. In one sense, the alienation of the gaze can, of course, be taken for granted. That we are shocked by the exposure to disabled life does happen all the time, and, as we have seen, does not have to be denied. However, whether or not this alienation is to be a fruitful one, whether or not we are shocked into a new perception and appreciation of our common human life with all its fragility and dependency, is a matter of the spirit that cries 'Fear not!' On the backdrop of the tendency to rationalize our deepseated anxiety in regard to disabled life in the way described above, the angelic 'fear not' is nothing less than a call for transformation. In the first instance it is a call for the transformation of our perception: of the disabled life and correspondingly our own life.

Hans Reinders points out the significance of a 'transformation experience'[37] that those who care for the disabled often undergo and testify. A peculiar transformation happens as the focus shifts from agonizing about the experience of a multiform hardship that is *thought* to await them to the *actual* experience of living with this particular human being and the multiform way in which they find themselves managing and growing in the process of facing difficulties as they actual encounter them.

Yet, as Reinders notes, the transformation at stake in taking on responsibility for dependent others would be seriously undermined by merely focusing on the shift from 'agonizing' to 'coping' or 'managing'. In an illuminating interpretation of Kenzaburo Oë's novel *A Personal Matter*,[38] Reinders marks out in his own way what we have called the moral challenge in the recognition of personhood and the angelic mission of dependent human life.

The novel portrays the conflict-driven evolution of the character of its principal protagonist, a man called Bird, after his wife gave birth to a disabled child. When agonizing about this fate, Bird is faced with what appear to be his 'two selves': the one self that urges him to distance himself from this 'monster' and escape into another life with another woman in another country – and the other, eventually triumphant, self that learns to embrace the responsibility for the entrusted life of his son and wife. The conflict between these two 'selves' resembles not so much 'choices' that Bird has to make as rather the challenge of coming to terms with the meaning of his own existence as a human being and the social commitments he has.

The image of the 'mirror' is a key in Oë's portrayal. When facing in the mirror his one self that is about to abandon child and wife for a 'better life' in Africa, Bird realizes that he could not wish or even stand to live with this self for the rest of his life. 'He pictured himself, having killed the baby, standing at her [the other women's] side . . . a sufficiently enticing prospect to Hell.'[39] Conversely and correspondingly, after his decision to return to the hospital and live his 'one' life with his wife and his son, Bird's discovering of himself anew is aptly pictured in another 'mirror image': 'Bird waited for the others to catch up and peered down at his son in the cradle of his wife's arms. He wanted to try reflecting his face in the baby's pupils. The mirror of the baby's eyes was a deep, lucid gray and it did begin to reflect an image, but one so excessively fine that Bird couldn't confirm his new face. As soon as he got home he would take a look in the mirror.'[40]

Oë's novel is a wonderfully hopeful story of recognizing personhood in the dual way we have suggested. It is a story in which the angelic mission of the disabled life eventually succeeds in reconciling those to whom it is entrusted with their own humanity. The new face that the main character is beginning to recognize in the mirror image of his son's pupils is no longer disfigured by fear but miraculously relaxed as it faces the future of a shared life with his disabled child.

Yet, whence precisely did this transformation come? Was it the father's own moral integrity that eventually managed to break through the cycle of temptation as to embrace the responsibility at hand? Or did the power to transform the father rather reside in the innocent and dependent life of the son that looked at him in a morally irresistible way? The subtlety in which the novel narrates the inner turmoil of the main character, including the abyss of hatred and despair in which he first loses himself, on the one hand, and the sober restraint from making much of the child's potential to charm him, suggests that it cannot be either of these alternatives. Neither the moral power of a Kantian sense of duty nor that of a Levinasian absolute claim in the countenance of the other renders the transformation of Bird intelligible as the novel portrays it. It appears more appropriate to say that the transformation is happening 'in between' the proponents whose individual capacities cannot account for it. The transformation emerges from the spiritual sphere of the *Mitwelt*, to use Plessner's term again. In this vein, the novel seems to suggest that the recognition of personhood, although its subject matter is precisely the shared human nature as such, is, at least at times, beyond the natural capacity of the individuals concerned.

To reconnect with our imagery of angelic mission for the recognition of personhood, we are led by the subtlety of the novel to understand better why the angel must not remain silent – a mere ontological phenomenon of meaning, as it were – but must really speak out and address those to whom she is sent. 'Fear not' is the message that needs to accompany the revelation of personhood as our shared life of human dependency, if that message is to be heard and embraced. In the language of the Christian tradition, in order to not cause fear, rejection and expulsion, the recognition of our kinship with the disabled cannot simply rest on the 'fact' of that which unites all humans as offspring of the same Ur-mother. While it presupposes and proclaims this common belonging of creaturely life, this reminder needs to be embedded in a wider recognition: the recognition of kinship with Christ who makes human creatures adopted sons and daughters of the Father. Whether conscious of it or not, every 'fear not' story, such as Oë's novel, implicitly feeds on the resurrection of the one *angelos tou theou* who did not recoil from having his wings clipped (Phil. 2.5–8) or from sharing the human life of fear and anxiety, yet was triumphantly raised from the dead to offer transformation and new life to those who recognize him and their own existence as human beings in the faces of his most dependent brothers and sisters (Mt. 25.40).

7 Conclusion

How do these considerations help us to reframe the question of 'personhood and the disabled'? The first point to observe is that recognition of personhood is *not* categorically different in the case of the disabled. It is not that a creative act of analogical imagination is needed if we are to count severely disabled members of the human race 'persons', whereas in the 'normal case' of encountering 'healthy human beings' it is just self-evident. The kind of recognition that draws on such a distinction would reflect an act of condescending inclusion as a result of a Promethean 'creating' of something ('rights of the disabled') that reflects the moral self-determination of those who bestow them instead of the nature of the bearer and his own inherent claims.

As I hope my considerations about the social and moral conditions of the language of 'person' have demonstrated, 'crisis' is the semantic seedbed of this notion. It is a moral notion with an originally emancipatory thrust. *Achtung: Person!* 'Person' functions as a caveat before it assumes any descriptive quality. The day-to-day usage of the notion in its seemingly harmless, i.e. conventional fashion ('33 persons will fit in this room when full'), feeds on this critical notion, and not the other way around.

This is why we can say that the life of a human being that we perceive as 'disabled' or 'retarded' is central instead of peripheral to the language-game of personhood. It is so because of its critical character – critical not in terms of the ontological question in regard to '*them*' (are they a fully human being?) but in terms of the challenge it puts to *us*: our moral response to its challenge on our expectations, career-planning, general outlook in life, both in individuals and in society at large. By insisting 'she is our child' when confronted with a devastating medical diagnosis of their new-born, or 'he is still a human being', when presented with a brutal murderer, parents and prison guards recognize themselves as responsible human beings ('persons') who are to care for someone whom they perceive as *entrusted* to them as family, no matter what qualities they might find lacking in the individual bearer of this nature.

Even though these parents or prison guards may tell us that 'they could not have reacted differently', we know they could well have. A phenomenologically saturated theological account of 'personhood' will not buy into the moral sentimentality of suggesting a sufficient 'natural' inclination of (human) nature to recognize (human) nature. Such instinctive certainty may be said to exist in and be sufficient for sheep and wolves. Human beings however, cannot recognize their kin, cannot recognize 'personhood', apart from having to *become* themselves what they recognize in others. The recognition of personhood is not a mere cognitive but, inevitably, a fully-fledged moral act. As such it may be described as an act of inclusion. Yet the one to be included is not the disabled but the 'moral agent' herself, who is to step into the realm of responsible humanity by recognizing her belonging with her kin.

We have described the process of recognition in terms of an angelic mission by which those who live amongst us with disabilities help those who are considered 'normal' to understand that all human life is inevitably marked by an elemental dependency. The courage and strength to accept this dependency both philosophically in terms of the anthropology we construe and practically in terms of the anthropology we embody will be aided by actually sharing our life with persons with disabilities. But given the cultural imperative to master and control, and the instinct it engenders to shy away from identifying ourselves with those who we think live at the margins, the 'fear not' of the angelic address must also be *proclaimed* as a matter of the public message of the Church.

We do not have to fear these angels with clipped wings or the message they bring us, as this message will liberate us from another, far more serious fear: the suspicion that we are only fully human – real persons, as it were – if we are not like them. This is a fear that must be exposed for what it is: a fear that cannot be lived by; not in the present as it makes us deny our elemental needs; and not in the future when the fantasy that we are in control of our lives will be ridiculed by the realities of old age. The fear that drives the discourse on 'human non-persons' as much as the 'inclusion' discourse eventually draws on the reality of death without hope. The reality that Christians believe in as one indestructibly marked by resurrection of the Crucified, has nothing to fear but everything to gain from the angelic mission of the disabled, as they challenge us to recognize our shared humanity of dependency and hope.

Notes

1. Peter Singer, *Practical Ethics* (Cambridge: Cambridge University Press, 1st edn, 1979); Peter Singer and Helga Kuhse, *Should the Baby Live? The Problem of Handicapped Infants* (Oxford: Oxford University Press, 1985); P. Singer, H. Kuhse, S. Buckle, K. Dawson, P. Kasimba (eds), *Embryo Experimentation* (Cambridge: Cambridge University Press, 1990); Norbert Hörster, *Neugeborene und das Recht auf Leben* (Frankfurt: Suhrkamp, 1995).
2. One example is H. Tristam Jr Engelhardt's *The Foundations of Bioethics* (New York and Oxford: Oxford University Press, 2nd edn, 1996). Engelhardt accepts the definition of persons in the vein of the secular discourse as those who can actively, by way of their 'agreement', establish a morally binding sphere of communication with others (p. 239), but adds that in regard to those who lack these capacities, societies can still establish practices that recognize them as persons 'for social considerations' (p. 147).
3. Helmuth Plessner, *Mit anderen Augen. Aspekte einer philosophischen Anthropologie* (Stuttgart: Reclam, 1982). Quotations from this work are given in my own translation, but the page references are to the German edition.
4. Plessner, *Mit anderen Augen*, p. 11.
5. Plessner, *Mit anderen Augen*, p. 14.
6. Plessner, *Mit anderen Augen*, p. 14.
7. Plessner, *Mit anderen Augen*, p. 13.

8. Plessner, *Mit anderen Augen*, p. 15.
9. Robert Spaemann, *Personen. Versuche über den Unterschied zwischen 'etwas' und 'jemand'* (Stuttgart: Klett-Cotta, 1996). As the English version is only forthcoming (Oxford: Oxford University Press, 2007), I am grateful to the editor and translator of the book, Oliver O'Donovan, for allowing me to use his translation in advance for the purpose of quotation. For obvious reasons the page numbers given refer to the German version.
10. Spaemann, *Personen*, p. 253.
11. Spaemann, *Personen*, p. 252.
12. Spaemann, *Personen*, p. 78.
13. Spaemann, *Personen*, p. 263.
14. Spaemann, *Personen*, p. 264.
15. Spaemann, *Personen*, p. 256.
16. Spaemann, *Personen*, pp. 81ff.
17. Spaemann, *Personen*, p. 156.
18. Spaemann, *Personen*, p. 261.
19. Spaemann, *Personen*, p. 260.
20. See Russell R. Reno, 'God or Mammon?', in C.E. Braaten and C. Seitz (eds), *I Am the Lord, Your God: Christian Reflections on the Ten Commandments* (Grand Rapids, MI: Eerdmans, 2005), pp. 218ff.
21. On the questions of defining power, see Bernd Wannenwetsch, '"What Is Man That You Are Mindful of Him?" Biotechnological Aspirations in the Light of Psalm Eight', Farmington papers E16 (2004).
22. In contrast, fathers in pagan Rome claimed such a 'defining' right as to decide whether or not to confer on the newborn the legal status of legitimate child and with it that of a human being. (Cf. Spaemann, *Personen*, p. 256.)
23. Dietrich Bonhoeffer, *Creation and Fall: A Theological Exposition of Genesis 1–3* (Augsburg: Fortress Press, 2004), p. 79. The idea of '*analogia relationis*' that Bonhoeffer introduced in such passages was taken up by Karl Barth in *Church Dogmatics*. 3.1. *The Doctrine of Creation* (ed. G.W. Bromiley and T.F. Torrance; Edinburgh: T&T Clark, 1958).
24. In an interview with the *New Yorker*, Singer is reported to have explained: 'Perhaps it's more difficult than I thought before, because it is different when it is your mother'. For a compelling reading of this inconsistency, see Richard John Neuhaus, 'A Curious Encounter with a Philosopher from Nowhere', *The Public Square, First Things*, 120 (February 2002), pp. 77–96.
25. Quoted in Rayna Rapp, *Testing Women, Testing the Fetus: The Social Impact of Amniocentesis in America* (New York: Routledge, 1999), p. 274.
26. Hannah Arendt, 'Es gibt nur ein Menschenrecht', in *Die Wandlung*, 4 (1949), pp. 754–70.
27. Michael Polanyi, *Personal Knowledge: Towards a Post-Critical Philosophy* (Chicago, IL: University of Chicago Press, 1958).
28. The family does not create our identities; God alone does that. The patristic church rightly rejected 'traducianism', the view that the

human soul was inherited like the body from ancestors, in favour of 'creationism', the view that each new human being was individually made by God. But the family is the first mediator of identity; it provides the structure within which the unique gift of self, which God bestows on every person, can be received. We are given our selves in and through God's self, but also in and through other human selves. From beginning to end we are in relation to God and our nearest neighbors; they are, though in different senses, our origin and destination (Gerit deKruijf, Gerd Höver, Oliver O'Donovan, Bernd Wannenwetsch, *The Freedom of the Family: An Ecumenical Contribution to a European Debate* [Münster: LIT-Verlag, 2007]).

29. Alasdair MacIntyre, *Dependent Rational Animals. Why Human Beings Need the Virtues* (The Paul Carus lectures; Chicago, IL: Open Court, 1999).

    Biblical language has depicted human life accordingly under leading metaphors such as (the notoriously dry organ) 'throat' as the original semantic background to what we translate as 'soul' (Hebrew: *nefesh*). Even Psalm 8 which is frequently cited as a testimony of the 'high anthropology' that comes with God's affectionate valuing of human life, uses terms for humanity that are telling: 'What is man that you are mindful of him, what is the son of man that you take note of him'? (v.4). The Hebrew word *enosh* does not simply mean 'man' but more specifically denotes his 'weakness' or 'frailty'. And the other word for 'man', employed in the second half of the verse, echoes the modesty of this nature: *ben adam* – taken from the dust to which he shall return.

30. Hans Reinders illustrates how deeply ingrained in our common perception is the refusal is to have one's horizon opened up to the truth that the disabled embody. He points out that when speaking of the 'value' or 'meaningfulness' of a disabled person's life, we are typically not speaking of what it means to *them* (which can be extremely difficult, if not impossible, to know), but what we assume it would mean to *us* if we had to live with a similar deficiency. 'If the panelist on the TV show states his view about a life with spina bifida, he is talking about his *own* life.' That is, he is fantastically comparing his actual life with what he presumes his life would be like under such very different circumstances. Hence the actual life of the disabled person remains outside the panelist's horizon, but this does not prevent him from assessing its value and meaning. To him we say: Judge not! (See Hans S. Reinders, *The Future of the Disabled in Liberal Society: An Ethical Analysis* [Notre Dame, IN: Notre Dame Press, 2000], p. 74.)

31. For an impressive testimony of this tendency in the sphere of reproductive ideologies, see Amy Laura Hall, *Conceiving Parenthood: The Protestant Spirit of Biotechnological Reproduction* (Grand Rapids, MI: Eerdmans, 2007).

32. As Karl Barth's angelology emphasized, the term 'angel' needs to be

understood not from the angle of ontology, from their very 'being' as inhabitants of the heavenly realm, but from their *function and ministry* of being God's *angeloi* (messengers) on earth (*Church Dogmatics*, 3.3: 369–410). Their business on earth is to reveal God's secrets to mankind, and although in the Bible most angels 'come down from heaven', there is enough scriptural warrant for special situations in which human beings themselves act as 'angels' precisely when being commissioned to reveal to their kin what is hidden but needs to be brought to light. An example is given in the narrative Acts 6.8–15 in which Stephen, when brought before the council and accused by false witnesses, is said to have had 'the face of an angel' (v.15) which proclaimed to the onlookers the truth they had wished to suppress.

33. The tendency to romanticize disabled individuals as innocent or sinless appears to be just another instance of the tendency to depersonalize them – as lacking constitutive features of personhood, in this case the capacity to make moral choices.
34. Plessner, *Mit anderen Augen*, p. 169.
35. Plessner, *Mit anderen Augen*, p. 172.
36. Plessner, *Mit anderen Augen*, p. 173.
37. Reinders, *The Future of the Disabled in Liberal Society*, pp. 175–92.
38. Kenzaburo Oë, *A Personal Matter* (trans. John Nathan; New York: Grove Weidenfeld, 1969).
39. Oë, *A Personal Matter*, p. 164.
40. Oë, *A Personal Matter*, p. 165.

12

Disability and the Quest for Perfection: A Moral and Theological Enquiry

Brent Waters

Over the past few years a number of remarkable advances have been made for treating various disabilities, and debilitating illnesses and injuries. For example, prosthetic arms and hands have been developed which enable recipients to put on their own socks, open jars and shave. With the aid of electrodes implanted in the brain, quadriplegics are able to move a cursor on a computer screen, as well as change channels and control the volume on a television monitor. Through electric stimulation of the brain, in tandem with physical therapy, some stroke patients have regained up to 40 per cent recovery of function in paralysed limbs. Experimental drugs have improved memory recall of Alzheimer's patients by up to 80 per cent for brief periods of time.

More effective methods of preventing the birth of individuals suffering severe disabilities or debilitating illnesses have also been developed. For instance, with a combination of *in vitro* fertilization (IVF) and pre-implantation genetic diagnosis (PGD), a widening range of genetic and chromosomal abnormalities can be identified. Embryos carrying these deleterious traits are not selected for implantation, thereby preventing needless suffering from such conditions as cystic fibrosis (CF), haemophilia, Down's syndrome and early-onset Alzheimer's disease. Foetal monitoring and testing techniques may also be used during pregnancy to achieve the same effect.

These therapeutic and preventive measures *in tandem* disclose a chilling perception of disability, raising in turn some disturbing implications for persons with disabilities. This is an admittedly a counterintuitive claim, for they appear to be based on opposing strategies. The success of the preventive measures is based on the destruction of so-called 'defective' embryos or foetuses. Although the purpose is to prevent suffering, there is nonetheless the troubling implication that it would have been better if people with preventable disabilities had not been born. Therapeutic advances, however, are seemingly countering this preventive strategy. With a combination of increasingly sophisticated prosthetics and more efficacious therapies, a perceived need for prevention will diminish.

It is mistaken, however, to perceive prevention and therapy in competitive terms, for they are more akin to complementary poles of a common strategy, namely, of enhancing human performance. Many, if not most, of the preventive and therapeutic measures currently being developed also have potential enhancement capabilities. Prosthetic limbs may someday produce better athletes. Neurolinks with computers may enable faster computation. Electrical stimulation might slow or prevent the loss of muscular strength associated with ageing. Drugs are already being used by individuals to improve memory and other cognitive functions. PGD can be used to select for desirable, rather than select against undesirable, traits, and perhaps in the future to insert specific genes. In short, the already fuzzy lines separating prevention, therapy and enhancement are growing increasingly indistinguishable.[1]

It is in the blurring of these lines that the *concept* of disability becomes increasingly problematic. Identifying a disability is not only diagnostic, but also entails judgements that are socially constructed. This social dimension is especially pertinent in determining relative degrees of impairment, and consequently how aggressively certain conditions should be treated or prevented. Making this determination means that some concept of what constitutes a normal person is operative, however vague or imprecise it might be. Moreover, enhancement presupposes that a standard of normality is not only a goal or measure but also an arbitrary limit to be surpassed. Yet this means that some notion or ideal of perfection, however vague or imprecise it might be, is also operative in order to determine the relative value of various enhancements. The idea of disability takes on the symbolic quality of a liability to be overcome in a larger quest for perfection, and hence persons with disabilities serve as stark reminders of the limits and conditions that this quest is endeavouring to eliminate.

In making this admittedly provocative claim, I am not impugning the motives of those developing more effective therapeutic and preventive techniques. I believe they are motivated largely by a compassionate commitment to alleviating suffering and improving the human condition. But in the absence of any clear understanding of what 'normality' and 'perfection' have come to mean in late modern societies, compassion can be easily distorted into a thin rationale for intolerance and discrimination. If true, then it is incumbent upon Christians to offer a counter discourse in which persons with disabilities do not embody a condition to be eliminated.

The purpose of this chapter is to make a case for the validity of this claim, and to offer a framework for a theological counter-discourse. The subsequent enquiry is conducted in three parts. In the first part, I trace the historical threads of a quest for perfection; in the second, I examine the relationship between liberal eugenics and quality control; and in the third, I propose some avenues that might be taken in developing a Christian counter-discourse.

The Pelagian legacy

Human perfection is a slippery ideal, for it has many different meanings. According to John Passmore, it can connote mastery of technical skills, subordination to God's will, attaining a natural end, freedom from moral defect, a metaphysical state, living an ideally perfect life, or becoming godlike.[2] Despite these various meanings, a concept of human perfection has nonetheless permeated Western civilization. Passmore contends that the origin of this concept has two principal strands. The first he associates with Athens. Plato knew that imitating the gods could not lead to perfection, for the conduct of the gods was morally defective. Humans, however, could perfect their social natures through orderly governance and civil discourse. It is in the life of the *polis* that men[3] learn how to become citizens. Hence his emphasis and reliance on education as the means of achieving the good contemplative life, for it is in contemplating 'the ordered glory of the universe' that the 'divine element' in every person can be perfected.[4] Aristotle builds upon this Platonic foundation by arguing that the soul is formed and perfected through practising the virtues. To be a just person, for instance, one must have a just soul, and in order to have a just soul one must live a life of justice. More importantly, the perfected soul is in communion with the divine.

According to Passmore, the legacy bequeathed by this Greek strand included a notion of 'metaphysical perfection', and the belief that at least some individuals were capable of attaining this quality. The principal means of such participation is knowledge gained through reason, and the number of individuals possessing this capacity is relatively few. Consequently, the ordering of the *polis* should support a contemplative life for this rare metaphysical elite, thereby reducing morality to a 'practical aim for the ordinary citizen'.[5]

Passmore traces the second strand back to Jerusalem. He asserts that early on Christianity rejected the idea of perfection, for both metaphysical and moral reasons. Humans simply lack an innate ability to become godlike, because sin has thoroughly corrupted their moral vision, and more importantly, disfigured their will. At best, sinners can only will good things badly. Moreover, this disordered desire has a corrosive effect upon both body and soul. Given the pervasiveness of sin, every person is, to varying degrees, physically, morally and spiritually disabled. Hence the need for the Incarnation, for it is only through divine grace that sinners might be healed. More importantly, since only Jesus Christ could be said to be the perfect human being, then any attempt to become like him through one's own initiative would be heretical.

It is precisely this heresy that Pelagius committed. For this morally rigorous monk the issue was clear: none other than Jesus himself had commanded his followers to be perfect as their heavenly Father is perfect.[6] Through their own effort, humans could become like Christ. In rejecting the doctrine of original sin that was passed on from generation to generation, the Pelagians insisted that a person was not born as a sinner. Humans imitated but did not inherit Adam's sin; a person could freely choose to become a sinner or a saint. Pelagius and his champions soon sparked the ire of St Augustine, who, in a series of polemical treatises, attacked the Pelagians for denying any need for divine grace, thereby stripping Christ of his redemptive purpose. Sin was not a matter of choice but a condition in which one was born, so redemption was not disciplined self-improvement but a lengthy process of healing that could never be completed this side of eternity. Following Augustine's lead, a series of church councils declared Pelagianism a heresy, and excommunicated its principal proponents.

Although Augustine won the initial battle, the outcome of the war is still in question. As Passmore contends, declaring Pelagius a heretic did not eradicate Pelagianism. A quest for spiritual perfection periodically surfaces and submerges in the Catholic mystical tradition, and determined efforts at becoming morally perfect have characterized a number of pietistic movements within Protestantism. Ironically, even latter-day Augustinians have proven to be susceptible. Cotton Mather, for instance, asserted that if humans could do all in their power to receive grace, then it was more likely that they would receive it. A person was not so disabled by sin that he could not will himself to be a recipient of God's power. For Passmore, this move proved portentous: 'How Augustine and Calvin must have shuddered in their graves! Thus it was that the United States gradually made its way, from being the most Calvinist, to being the most Pelagian of Christian nations.'[7] Later, F.R. Tennant would more ingeniously argue that although humans cannot achieve any absolute perfection of their own volition, they can nonetheless achieve a standard that is applicable and attainable.

However interesting this long intramural theological debate might be, what Passmore finds more troubling is the modern, secular manifestations of Pelagianism. Beginning with the Renaissance, the emphasis shifted from perfecting individuals to perfecting humankind. Perfection had nothing to do with metaphysics, but should focus on 'task-performance'.[8] The issue at hand is moral rather than spiritual, for it is through the specialized and collective efforts of individuals that humankind perfects itself. This task is taken up with a vengeance in the Enlightenment, during which education replaces grace as the most promising route for improving the human condition. Ignorance, not sin, is the enemy of progress, and it can be conquered with knowledge.

This confidence in human intellect spawned what may be described as a series of modern projects in social, political and biological engineering throughout much of the nineteenth and twentieth centuries. Good planning and efficient execution accompanied with scientific precision would make utopian dreams come true. Activists, scientists, social workers and politicians joined ranks in reforming education, criminal justice, health-care and the workplace. If humankind could not be perfected through legislative reforms, then at least such disabilities as ignorance, crime, ill-health and poverty could be ameliorated if not eventually eliminated. The notion of progress *per se* was never an issue; the debate was whether the engine of progress was driven best by liberal democratic and capitalistic principles or socialist and collectivist strategies.

Despite this ambitious social and political agenda, some individuals and groups were seemingly unable to benefit from these reforms. Some were incapable of learning, while others were habitual criminals; some suffered chronic and incurable diseases, while others would never be able to support themselves. The rediscovery of Mendel's studies in genetic inheritance, however, seemingly offered a solution to this perplexing dilemma: if certain individuals could not be helped, at least the suffering of future generations could be prevented by removing their seed from the gene-pool. The eugenics movement was founded on the assumption that the most pressing social and political problems were perpetuated primarily by certain distinct classes and categories. If these people were prevented from reproducing – with or without their consent – such problems as crime and poverty would virtually disappear within a generation or two.

Passmore concludes his ambitious survey with the observation that although the two world wars diminished interest in the idea of human perfection, it did not go away. Indeed, as the twentieth century was drawing to a close, the prospect of genetic engineering, and biotechnology more broadly, was rejuvenating renewed visions of perfectibility, though often cloaked by the rhetoric of therapeutic enhancement.

Although Passmore's book is dated, and his sweeping generalizations often occlude as much as they enlighten, he nonetheless offers this salient insight: various notions of human perfectibility have waxed and waned within the history of Western civilization. Yet whenever a particular vision of perfection seizes the public imagination, it is accompanied by a corresponding decline in tolerating perceived deviations. Moreover, these perceptions are based on both purportedly diagnostic *and* normative criteria. To identify a disability is not only to acknowledge abnormality, but also to pass judgement. Disability not only connotes difference but also inferiority; a condition that needs to be either rectified or prevented, and therefore eventually eliminated.

Consequently, disability is simultaneously a biological, social and political 'problem', both in terms of perception and remedy. This is especially apparent when eugenics is proffered as the best solution for such problems as public health, poverty and crime. We need not impugn the motives of the leaders of this ill-fated movement, for their rhetoric appealed to such seemingly worthy goals as preventing suffering and improving health. What it does exemplify is that good motives are not enough, especially when they are linked with bad science. Such naïvety rationalizes a distorted moral vision in which the disabled *person* comes to be seen as only a disability, a moral blind-spot that was shared equally by ideologies of both the left and right.[9] And as a depersonalized disability it stood as a symbolic obstacle to be eliminated in the quest of perfecting, or at least substantially improving, the quality of human life.

Liberal eugenics and quality control

Is Passmore's allusion to a renewed interest in human perfectibility with the prospect of genetic engineering and biotechnology a warning that Western civilization is about to enter a new, more horrific round of eugenics? Perhaps, but it seems unlikely that wideranging policies similar to those imposed by both democratic and totalitarian regimes in the previous century are being contemplated. This does not mean, however, that eugenics has been expunged from the Western imagination. Rather, it is resurfacing in late liberal societies as part of a broader movement of exerting greater quality-control over human life and individual lives.

To assert that the concept of quality-control has become a formative force in contemporary medicine is merely to state the obvious. At the end of life, for example, greater qualitative control is exerted, purportedly, through improved palliative care, euthanasia and assisted suicide. The advent of regenerative medicine is not only promising greater longevity, but also stemming the loss of physical strength and cognitive functions associated with ageing. In both instances, medical treatments incorporate what may be characterized as collapsing enhancement *and* preventive measures into a singular health-care regimen. Giving a dying patient greater control over the time and means of her death is an enhancement that prevents a type of death she determines to be unwanted or undignified. Preventing a decline in physical and cognitive performance is also enhancing the life of an ageing patient.

It is with the beginning of life, however, where combining prevention and enhancement is most problematic, for it requires exerting greater control over the characteristics of offspring rather than the quality of one's own life, or that of another person in one's custodial care.[10] Reproductive technology is being used increasingly not only to assist procreation but to control the qualitative outcome of the reproductive process. As mentioned above, this often means using PGD to prevent the birth of a child with a disabling disease such as cystic fibrosis or Down's syndrome. The same technique is now also being used to select for desirable traits, such as a gene that can be used to treat a sick sibling, and for 'social reasons' such as sex-selection.

Selecting for or against certain genetic traits is a form of eugenics, but what makes this particular process 'liberal'? It may be designated as liberal because exerting quality-control presumably enhances the freedom of individuals in pursuing their reproductive interests. Within the tenets of 'procreative liberty', each individual has a fundamental right 'either to have children or to avoid having them'.[11] Individuals who want children should also have unrestricted access to technologies that will help them fulfil this desire. Since all competent adults have a right to obtain a child, then they also the right to obtain a desirable one. Consequently, quality-control is 'a core part of procreative liberty and should be respected as such. If it is legitimate for parents to want healthy children, then it should be legitimate for them to use both negative and positive techniques to achieve that end'.[12] Moreover, if genetic enhancement of embryos proves feasible, there should be no qualms in pursuing it, for such an intervention is little different than parents enhancing their children by enrolling them in prestigious schools or rigorous athletic programmes.[13]

If the presuppositions of procreative liberty are accepted, then exerting quality-control over the characteristics of offspring plays a significant role in enlarging the scope of individual freedom which is defined almost exclusively in terms of the absence of external constraints against one's will. If a person wills to have a child, then he should have the right to want and obtain a desirable child. Placing restraints upon quality-control would, moreover, diminish parental satisfaction, thereby diminishing their freedom as well. Most importantly, this expanded range of reproductive freedom is achieved without threatening the welfare of others, for in selecting for and against specific embryos no person is harmed in the process.

It may be objected, however, that if such quality control becomes practised widely it will produce a prejudicial climate against persons with disabilities. They will come to be regarded as a class of unwanted persons: products of imperfect preventive techniques or irresponsible parents. This will especially become the case if economic incentives or disincentives are enacted that encourage potential parents 'at risk' to take advantage of various quality-control techniques. Although persons with disabilities are not harmed directly by the use of PGD, their interests will nonetheless be harmed indirectly as its use, or the use of similar testing techniques, become widespread. Consequently, in order to protect persons with disabilities, legislation should be enacted which restricts the range of conditions that can be tested for and selected against.

It is not likely that such legislation will be forthcoming, for its effectiveness would require changes in laws governing therapeutic abortions, as well as extensive alterations in how IVF is often performed. Moreover, even if sufficient political will could be created for preventing the practice of selecting against embryos (and foetuses) carrying certain deleterious traits, such a prohibition would not be extended to selecting for embryos (and foetuses) that are deemed desirable for either social or therapeutic reasons. Prevention and enhancement have already become so intertwined in reproductive medicine that any attempt to unravel them would require recourse to illiberal restrictions on individual freedom, the very antithesis for which the development of reproductive technology was designed to promote. In short, no compelling objection against liberal eugenics can be raised *on liberal grounds*.

This can be seen in Jürgen Habermas's failed attempt to argue against the prospect of parents modifying the genes of their offspring.[14] Habermas presumes that he can offer an irrefutable argument against genetic enhancement by appealing to the bedrock liberal principle of autonomy. The autonomy of the purported beneficiaries is violated because they are unable to give their free and informed consent to having their genes altered. A 'previously unheard-of relationship arises when a person makes an irreversible decision about the natural traits of another person'.[15] This ability to manipulate offends our moral sensibilities by corrupting the delicate balance of the parent–child relationship. Parents should not have the power to select their children in the same way they choose other goods and commodities.

How does Habermas propose to safeguard our wounded sensibilities against 'the obscenity of this reifying practice' that is forming 'a society which is ready to swap sensitivity regarding the normative and natural foundations of its existence for the narcissistic indulgence of our own preferences'?[16] Liberal eugenics is immoral for the simple reason that it disables enhanced offspring from exercising their autonomous self-actualization. Genetic enhancement is wrong, because it violates the premier liberal conviction that persons should not be forced to pursue lives they do not choose. Consequently, a person should have the right not to inherit a modified genome. In addition, liberal eugenics erodes the social relationships comprising civil society, for the fundamental equality underlying the principle of autonomy is derived from a process of lifelong socialization. Involuntary genetic alteration cannot be practised fairly in a liberal democratic society without destroying a foundational commitment to both autonomy and equality.

Habermas is confident that his appeal to autonomy offers an impregnable bulwark that simultaneously strips liberal eugenics of any plausible justification, while safeguarding the central liberal tenet of freedom. He accomplishes this task by avoiding any mention of embryos as persons, while preserving the moral primacy of personhood. This tactic allows Habermas to undercut what he believes is the only possible justification for genetic enhancement, namely, that it is a variation of lifelong socialization; that 'there is no great difference between eugenics and education'.[17] But eugenic interventions violate the social relationships within which autonomy is formed; persons are free to modify themselves, but they have no right to modify others who cannot freely grant or withhold their consent.

Given the presumed strength of his argument, Habermas is perplexed why most of his critics do not engage it directly but merely bypass it altogether.[18] What he fails to recognize is that he need not be engaged directly, because the autonomy he is endeavouring to defend has already been negated by the technologies that make so-called liberal eugenics possible. Embodied personhood is no longer definitive, but an unfortunate constraint against the will, and therefore a limit to be overcome rather than honoured. In its emphasis on an assertive will, late liberal social and political thought has merely exacerbated the problematic status of the body. Habermas's so-called natural and normative foundations that are being attacked by liberal eugenics cannot be defended, for they have already been worn away by the acidic assault of the disembodied will. Many individuals may be offended by liberal eugenics, but their offence is emotive rather than rational or normative. What Habermas fails to recognize is that late liberal society is no longer comprised of embodied persons, but populated by wills that happen to have bodies.

Habermas is rightfully offended that procreation and childrearing has shifted from the natural and the social 'to the realm of artefacts and their production'.[19] His offence exemplifies why his bulwark of autonomy can be easily bypassed, for late liberal society has also made a postmodern shift in which artifice is displacing nature as the primary metaphor for social and political ordering.[20] Both the persons *and* associations comprising civil society are malleable constructs. When children are made rather than begotten then birth is no longer a given event, but the outcome of a reproductive will, and the associations which in turn socialize them are products of a fickle political will. The wilful construction of human life and lives is both a biological and social project, but the two categories are collapsed into a singular enterprise of enhancement. Moreover, since the limits of such construction are, in principle, infinite, then so too are the enhancement possibilities. The goal is not simply to make humans better, but to produce beings that are better than human. Although it is tempting to dismiss the hyperbole of posthuman rhetoric, it nonetheless designates a war being waged to free the will from physical and temporal constraints.[21] In this respect, all humans are disabled by the will's inefficient prosthetic (the body) that has been bequeathed by biological and cultural evolution. This means, however, that persons with disabilities are rendered invisible, for they are no longer perceived as embodying conditions to be treated or prevented, but are simply part of the morass of inferior material with which to work in constructing the superior posthuman.

Embodied community

If the preceding analysis is at all correct, then it is incumbent upon Christian theology to develop a counter-discourse. As Habermas demonstrates, the morality of liberal eugenics cannot be challenged on its own philosophical terms. The issue at stake, however, is not prevention or enhancement *per se*, but the notions of personhood and perfection that inform how technologies are and should be used to prevent or enhance. Late liberalism has largely distilled the person down to a self-conscious and assertive will. Consequently, the physical body always serves to constrain persons in a wilful pursuit of their respective interests. The severity of a disability is determined in relation to the extent that it restrains individuals from identifying and pursuing these interests, as well as informing what preventive or therapeutic treatments should be used in response. Perfection, then, is presumably achieved in the absence of any external constraints upon the will.

For Christians, in contrast, human perfection is Jesus Christ, and a disability is any condition that prevents one from being made perfect in Christ. This is merely to restate the traditional claim of Christian theology that anthropology *is* Christology, but it must be developed and expressed in a manner that resists the corrupting tendencies of Greek and Pelagian thought as traced out by Passmore. In this final section I will not indicate what the content of such a theological anthropology might entail, for such an account would require detailed doctrinal analysis and development that is beyond the scope of this chapter. Rather, I will suggest some avenues such an enquiry might take, as well as some pitfalls to be avoided.[22]

If Christ is perfection, then sin is the disability that prevents one from becoming Christ-like. Consequently, all humans may be said to be disabled, for all humans are also sinners. Making this claim does not require an appeal to a historical Fall, or transmission of original sin from one generation to the next. Rather, it acknowledges that certain capacities which enable human survival may also disable human flourishing within a matrix of changing social circumstances over time. This tension can be seen in both biological and cultural evolution. The body's ability to store fat, for example, enables survival, particularly when food is scarce. But the same capacity can become disabling for some, if not many, individuals within a sedentary culture in which food is plentiful and badly prepared. This does not imply that the sin of gluttony represents a fall from some prior pristine or perfect diet, or that a moral defect is being passed on through the genes by individuals with a proclivity for obesity. It instead serves as a reminder that a particular disability is *both* diagnostic and contextual. The disabling diseases associated with obesity are partly the result of genetic predispositions, but they are also partly triggered and exacerbated by particular cultural circumstances. To claim that obesity is solely a medical problem which could, at least in principle, be prevented, and eventually eliminated through screening and testing, naïvely privileges biology over social and political ordering. Reordering the fabric of daily life in ways that promote better diets and exercise, for example, is arguably a superior preventive measure than using PGD (or similar techniques) to prevent the birth of persons with a particular genetic predisposition.

Although it is important to emphasize that all humans are disabled by sin, care must be taken regarding the tone of the emphasis. To say that all are disabled by sin does not entitle us to say that all of us are disabled persons. To make this patronizing move is tantamount to making persons with disabilities invisible. The ubiquity of sin is manifested in particular, rather than common, ways, as expressed through a varying range of specific needs. In short, persons with disabilities need to be accommodated, and in some instances assisted, if they are to participate more fully in civil community. In this respect, appropriate therapy, as opposed to prevention or enhancement, expresses a love for neighbour. The provision of sophisticated prosthetic limbs, for instance, is no small benefit in a society that values mobility. If therapy is to be a loving response to a particular neighbour in need, however, then it must also be acknowledged that not all therapeutic applications are always appropriate. A cochlear implant, for example, is not necessarily needed by every deaf person. Stressing the ubiquity of sin serves as a reminder that the natural and social affinities that bind humans together are fragile, and without proper attention they corrode and become weakened.

If these affinities are to be strengthened, then the emphasis on sin must be accompanied by an equally strong emphasis on grace, especially its gratuitous and gifted character. Pelagius's fateful error was to believe that perfection is a reward to be earned. If true, then there is both an implicit imperative *and* entitlement to seek perfection. And exercising the right to become perfect may entail the necessity of preventing and eliminating the imperfect. Hence, the cold logic of eugenics and posthuman construction that seemingly can only be accepted and not resisted by late liberal societies, for it is not coincidental that it is being played-out with a peculiar urgency in what Passmore identifies as the 'most Pelagian of Christian nations' (the USA).

In formulating a counter-discourse capable of resisting this logic, Christians must insist that whatever perfection may be said to exist, it is given and received as a gift. It is through the gift and grace of the Incarnation that God and creation are bound together. And derivatively, it is through grace that the natural and social affinities that bind humans together are sustained and strengthened through an economy of the giving and receiving of gifts.[23] Accommodating the needs of persons with disabilities is based neither on entitlement nor benefaction, but on the recognition that in keeping company with each other, each is graced by the gift of the other. However disabling a condition might be, the relationship is always reciprocal,[24] so that persons with disabilities remain members of the civil community with particular needs, rather than objects of care and pity. May not the scholar disabled by critical cynicism, for instance, be graced by the gift of simplicity offered him by the Down's syndrome person?

Worship may prove to be the most vital resource in formulating this counter-discourse,[25] for it is in worship that the gifted quality of grace is embodied. When Christians gather to worship they do so not only as an association of believers but as the body of Christ. Following Paul's imagery,[26] there is a relationship between the parts and the whole: the whole orders and affects the parts, while the parts, in turn, affect the whole. We may say, then, that sin disables the body as a whole, and that the disabilities of particular members affect and form the body in its wholeness. We may also say that grace is given as a gift to the body as a whole, enabling the body in turn to receive and be graced by the gifts given to it by persons with disabilities. Moreover, the proper response to the giving and receiving of gifts is gratitude. This does not imply that disabilities are good, and should inspire a response of thankfulness. Indeed, Christians properly rejoice when the lame walk and blind see. Rather, the imagery of a disabled *and* graceful body suggests that there are alternative perceptions of disability other than an unmitigated tragedy that should be prevented by any means possible. There is no reason to presume that persons with disabilities are somehow disabled from full fellowship in the body of Christ.

The embodied imagery of worship is especially amplified in the Eucharist. Christians gather at the Lord's Table to share a common meal. Through a single loaf and chalice they receive the gift of Christ's body and blood, and as Christ's body disabled by sin are graced by this gift. It is also through these common and finite elements that they are joined together with other Christians separated by time and space into Christ's body. It is through this sacrament that we may say that the whole of this body is constituted by and embraces the particular parts, while each part may also be said to contain the whole body within it. Does this not suggest, then, that when I kneel to receive the bread and wine, I do so not only for myself, but also for sisters and brothers in Christ who are unable to do so? Does it not also suggest that this simple act is a gift that has graced both giver and recipient? And might this exchange of gifts offer more expansive imagery for ordering our social and political associations in ways that acknowledge that persons with disabilities have particular needs that require accommodation, but these needs have not disabled them to any greater extent than any other sinner?

I want to be clear that this act of kneeling for others who are unable to do so is not motivated by any sense of pity or eviscerated charity. It does not really matter who is kneeling and who is not, in receiving the grace given in the gift of Christ's body and blood – for all are one in Christ. The sacrament is received by individuals but given to the community. When I kneel to receive the sacrament it is not my sisters and brothers who are unable to kneel who benefit from my act. Rather, I am the beneficiary for I am reminded that I am bound together in Christ with persons who are not like me, and who in turn I am forbidden to exclude or pity because of the dissimilarity. When I kneel when others are unable to do so, I am (or should be) reminded that since we are one in Christ what I do is not isolated from what others do or are unable to do, and conversely what others do or are unable to do affect me. The condition of any part affects the whole, and the whole in turn is embodied in each part.

Consequently, when I kneel when others are unable to do so, I am (or should be) reminded that the hope which binds the body of Christ together is eschatological. It is in the eternal life of God in Christ that sinners are redeemed and all are made whole. In this respect, the resurrection of the body serves as a powerful reminder that it is as embodied creatures and not disembodied souls (now read by late moderns as unfettered wills) that are in communion with and within the eternal life of the triune God. Furthermore, this hope reminds Christians that their anthropology is also eschatological. Whatever perfection they might enjoy awaits them as a destiny and gratuitous gift given by God in the fullness of time. Thus any inclination of achieving perfection on our own terms is to be rejected as a false hope and counsel of despair, for, as in all Pelagian delusions, it ultimately requires a loathing of our status as finite and vulnerable creatures. And such loathing results inevitably in an intolerance of persons who cannot be perfected this side of the eschaton. But if perfection is in Christ and Christ alone, then no human is given the authority to make this judgement. In admittedly crude terms, when I kneel when others are unable to do so, I am also reminded that I am forbidden to regard them either as mistakes to be corrected or as individuals who should have been prevented from being born. I am to behold them as they are: as my sisters and brothers in Christ.[27]

It may be objected that this sacramental imagery is inconsequential, because it is symbolic and therefore has nothing substantive to offer in respect to how preventive, therapeutic and enhancement techniques should be used. My reply is that yes, the imagery is symbolic, but is it not symbols, such as disability and perfection, that shape, for good or ill, our moral vision and deeds?

Notes

1. See Robert Song, *Human Genetics: Fabricating the Future* (London: Darton, Longman & Todd, 2002), pp. 58–78.
2. See John Passmore, *The Perfectibility of Man* (New York: Scribner's, 1970), p. 17.
3. The public life of the *polis* is gender-specific: see Jean Bethke Elshtain, *Public Man, Private Woman: Women in Social and Political Thought* (Princeton, NJ: Princeton University Press, 1993), pp. 19–54.
4. See Passmore, *The Perfectibility of Man*, pp. 43–45.
5. See Passmore, *The Perfectibility of Man*, p. 67.
6. Mt. 5.48.
7. Passmore, *The Perfectibility of Man*, p. 115; cf. Michael Hanby, *Augustine and Modernity* (London and New York: Routledge, 2003), pp. 106–33.
8. Passmore, *The Perfectibility of Man*, p. 152.

9. It should be noted that the majority of involuntary sterilizations in the USA were performed in California, while the majority in Europe were performed in Sweden, prior to the rise of the Third Reich.
10. For the sake of the following analysis, I am assuming that human embryos are not persons, and their wilful destruction is not wrong or immoral.
11. See John A. Robertson, *Children of Choice: Freedom and the New Reproductive Technologies* (Princeton, NJ: Princeton University Press, 1994), p. 22.
12. Robertson, *Children of Choice*, p. 172.
13. I examine and critique Robertson's account of procreative liberty more extensively in *Reproductive Technology: Towards a Theology of Procreative Stewardship* (London: Darton, Longman & Todd, 2001), pp. 19–31; see also Gilbert C. Meilaender, *Body, Soul, and Bioethics* (Notre Dame, IN: University of Notre Dame Press, 1995), pp. 61–88.
14. For a more extensive critique of Habermas, see Brent Waters, *From Human to Posthuman: Christian Theology and Technology in a Postmodern World* (Aldershot: Ashgate, 2006), pp. 37–40.
15. Jürgen Habermas, *The Future of Human Nature* (Cambridge: Polity Press, 2003), p. 14.
16. Habermas, *The Future of Human Nature*, p. 20.
17. Habermas, *The Future of Human Nature*, p. 49.
18. See Habermas, *The Future of Human Nature*, pp. 75–100.
19. Habermas, *The Future of Human Nature*, p. 12.
20. See Richard J. Bernstein, *The New Constellation: The Ethical–Political Horizons of Modernity/Postmodernity* (Cambridge, MA: MIT Press, 1991).
21. See, e.g., N. Katherine Hayles, *How We Became Posthuman: Virtual Bodies in Cybernetics, Literature, and Informatics* (Chicago, IL: University of Chicago Press, 1999), and Elaine L. Graham, *Representations of the Post/Human: Monsters, Aliens and Others in Popular Culture* (Manchester: Manchester University Press, 2002).
22. For a more in-depth enquiry into this topic, see my article 'What Is Christian about Christian Bioethics', *Christian Bioethics*, 11.3 (December 2005), pp. 281–95.
23. See Stephen H. Webb, *The Gifting God: A Trinitarian Ethics of Excess* (New York and Oxford: Oxford University Press, 1996).
24. In some instances this reciprocity can only be expressed symbolically.
25. See Bernd Wannenwetsch, *Political Worship: Ethics for Christian Citizens* (Oxford and New York: Oxford University Press, 2004).
26. See 1 Cor. 12.12–31.
27. Nothing I have written in the preceding paragraphs should be construed as being anti-therapeutic. There is nothing inherently wrong with humans using medicine to heal and restore health.

13

The Broken Body and the Disabled Body: Reflections on Disability and the Objects of Medicine[1]

Jeffrey P. Bishop

Introduction

I want to believe that no doctor ever goes into health-care to have mastery over other people. In addition, I have also noticed that every medical student application form seems to say that the reason the prospective student is choosing the field of medicine is to help people. That medical schools might beat that out of its students is another essay entirely.

So how is it that the disability rights critique of medicine claims that medicine seeks to claim power over the bodies of those deemed abnormal? No recent critique has been so scathing of biomedicine than that of those scholars – and non-scholars,[2] for that matter, in the form of people resisting categorization – who have held a mirror to biomedicine's face, allowing, or forcing it to see itself. Biomedicine has not wanted to see that reflection for a myriad of reasons, not least of which is that every doctor knows that she went into medicine to help people. That she might actually be hurting or harming people in and through the very means by which she claims to be helping people is unthinkable – it is too much to handle. The physician's self-image looks so incredibly different to the image that the disability critique has reflected back to us.

I will show that in the very heart of modern medicine, and in the creation of its knowledge structures, there is something that results in the necessary objectification of the body – that is to say, an objectification that is necessary if modern science is to be employed to achieve the goals that it and society articulates. In this sense, Eric Krakauer's statement is apposites: Modern biomedicine is the 'standard bearer of western metaphysics'.[3] The most important mode of being for biomedicine is the mode of objectness. That medicine makes its objects is seen most acutely in the disabled body, in the body that does not function – where function is not just physiological function, but now societal function. That is to say, the body of one deemed disabled is made known as object of medical enquiry because it does not function.

In the first part of this chapter, I will show that this necessary objectification inevitably results in something like eugenics. In the second part, I will look to ethics, particularly the ethics of Emmanuel Levinas and the call of the other for the possible source of medicine's salvation. Disability – the malfunctioning body – appears as other in a way most acute to modern medicine, precisely because it fails as object when it does not function, particularly at the level of social production. Yet, I will show how Levinasian ethics also fail to solve the problem. Moreover, I will go on to show how technology so removes us from the other that it becomes difficult to allow the other to be. I will then turn to how Jean-Luc Marion's phenomenology might offer us recourse for medicine's salvation, toward a medicine that will be constructed by the broken body itself, which itself elicits a desire not to manipulate or master the other's body but to care for and meet the needs of this particular other. In short, I will claim that the disabled body is a body that fails to meet the demands of the idealized god of modern society – the productive body – and appeal to what is given to medicine in the broken body, an icon for medicine. I will conclude with a theological reflection on the practice of liturgy as a practice in openness to the other.

Part 1 The true, or scientific being

The subject and object of modern medicine

When doing medicine, the story goes, one must be objective. What do we doctors usually mean by this? First, on the side of the patient – that is to say, the side of the object – it typically means that the physician has a body of knowledge that stands outside the patient's cultural, spiritual and/or personal belief systems. Typically in contemporary Western society, when we have a medical problem, we want someone who can see through all the cultural interpretations ascribed by the patient. We want to get around the gap between perception and reality; we want someone who can abstract from the patient's cultural, spiritual and personal beliefs, to get at the 'real' problem. In addition, people generally want someone who can set aside the fears and anxieties about what this health blemish might mean, and thus to be able to give him the 'true' reading of the problem. In other words, one generally wants someone to read the problem from outside one's own messy cultural interpretations or cultural and spiritual perspectives, as well as the isolated perspective of one's own reading of one's body. And if this is not what patients want, it is the story that doctors are told that the patient wants.

Secondly, on the side of the physician – the subject – the call for objectivity can mean that one wants a physician who will look at the problem impersonally and reasonably. The old adage is that one should never take a family member as a patient, because one's own feelings can get in the way of cool reason. Reason is clouded by feelings and fears as it is, and thus the quintessential physician might succumb to emotionality, missing the diagnosis or botching the treatment. The physician's categorical knowledge is given privilege because her perspective is objective, outside time, place and circumstances, oddly, outside her own personal perspective. The physician is the subject that transcends her own subjectivity to see that which is truly patient to her agency. The physician is the subject that perceives the object in its raw facticity. The disease can be known as the object of the physician's true gaze. Or, at least, so the story goes.

Thirdly, when we say the physician must be objective, we can also mean that we want someone who will treat the body as an object of enquiry or manipulation. In some sense it is necessary to the practice of medicine that the body emerges as object, or shall I say, the body is reduced to object/machine by the subject. In some ways, we want the physician to shift horizons, to the horizon of being-object. That medicine does this readily is not merely a problem with Western medicine, but it is something that we the potential patients want. For example, when I speak to groups I often get a response from potential patients to the effect of, 'I just want the technically best physician/surgeon I can get. I do not care about bedside manner.' Within modern medicine, which is a product of Western culture, there is a real sense that the body is a machine acting on certain principles not known to the untrained. Thus, we want a good mechanic, someone who understands the body in the immanent realm of cause and effect, efficient causality.

It is here in our understandings of causality that medicine buys into the metaphysics of modern natural sciences,[4] or becomes the 'standard bearer of western metaphysics'.[5] With the rise of modern science, the older Aristotelian notion of causality is abandoned, and in so far as modern medicine becomes scientific, it adopts the metaphysics of the early modern natural sciences. While many in the natural sciences are beginning to rethink the role of metaphysics, I would argue that modern medicine remains within a modern metaphysics, dropping formal or final causality as speculation, leaving only efficient causality assessed and known through its effects by empirical methods, and placed within rational models. Another way of putting this is that modern medicine seeks to understand function, not purpose – mechanism, not meaning.

Thus, for a scientific medicine, the body becomes an object that operates on the principles of mechanism, which can be understood with reason. Thus, medicine, in so far as it seeks to be scientific, looks only at function within the realm of efficient causes. The function of the body-machine is lifted out of the messiness of human purpose.

Thus, for all the talk about the face of the other, when someone goes to the doctor with a facial blemish, the doctor does not listen for the call of the other, but perceives an object: the blemish on the face.[6] She looks at the skin, or the chin or the nose, in isolation. It does not matter which skin, chin or nose appears before her; it does not matter what the person whose chin it is believes about his blemish. The object, namely the nose or wart, or malfunctioning bit, is subjected to the physician's categorical and universal knowledge.

If the patient has feelings, then these likewise become objects to be manipulated by other subjects in the field of psychiatry. If the patient's social circumstances get in the way of the physician doing her job, then the patient is referred to a social worker who treats her relationships as objects to be manipulated. The patient is fully inscribed within a new polis where surveillance of health is carried out by various experts who objectify the patient in all of her modes of being – biological, psychological, social, cultural, political and, most recently, spiritual.

Finally, there is also a social and political dimension to the practice of medicine itself, which comes to ensure that physicians behave in this way, referring the patient to those who have the expertise to fix the patient within a set of expert discourses. And to extend the surveillance further, the physician is judged by her ability to perceive the patient as object. Within the practice of modern medicine, it is socially and politically necessary to treat the body as machine. I am judged by my peers on how well I know the body as machine, with all the bits, including the emotional, social, psychological bits, monitored by experts who subject the patient to a gaze.

Now, without being too judgemental of my profession, this bracketing of the personal perspective – her own and her patient's – has become necessary. She sees and knows different things about the patient by shifting the horizon from the phenomenal experience of the person to the horizon of beings, of objects. The chin or nose, or lack of proper function, comes to be assessed as object of investigation and the object of manipulation. If one is to fix the broken machine, to make the machine more efficient for living, she must shift to this horizon. The best physician then is not one who always keeps the personal view – her own or the patient's – out of the picture, but who hides the fact of her action of shifting horizons.

The patient is truly patient here. He is the passive object of the physician's agency. The physician, the subject in the subject–object relationship, subjects the object to her intellectual pursuits. In a way, the subjecting of the object to the gaze is pre-decided, prescribed, as Krakauer puts it.[7] The physician, the one that is well liked by her patient, is the one who masks what she is actually doing, subjecting, manipulating and mastering her object without regard to purpose. And that subjecting, manipulating and mastering is a good thing; or, at least, so the story goes.

Yet I would argue that the doctor, in forgoing purpose, merely shifts to the patient as object within the larger horizon of society, which can itself be moulded and manipulated within a society. And when one moves medicine from the individual subject–object relationship to medicine as discipline, disciplining the body, one is left with a medicine inscribed within society's functions, and in modern Western societies that function is work: productivity within a capitalist economic framework. That is to say, the activity which transcends the material objects of society is production in a capitalist society.

The subject–object and political economy

While medicine attempts to make the body a medical fact, an object, it is instead inscribed within political and historical relationships. It is a body with a political function. The functioning body is the body that fits into the political landscape. Foucault makes this point when he lists the various ways in which the French

Revolution began to speak of the various ill-healths of the body-politic. He would later name this concern for governing the body-politic as biopolitics.[8] The body becomes a bit, a machine, in the larger machine of society. The bodies of the body-politic must function properly for society to function properly. The ill-health of one part becomes ill-health in the body politic. Efficient causality extends beyond the mere object of the body and into the series of causes and effects within the social arena. The body is the object of relationships, driven by what I would call the new queen of the sciences, namely economics.[9] For political economy now orders our relationships, what gets researched, who gets to do the research, what counts as important knowledge. All is defined by who gets the money to do the research. The utilitarian scheme of efficient causality inscribes the body in its social contexts, and the disabled body, from the perspective of modern medicine, is without utility in efficient production.

For Foucault, this idea of biopolitics is no longer the oppressive power of society to define disease, death and the abnormal, but it would become, in his later works, the position of society to wield power, creating the circumstances for new vitalities and freedoms. But the important thing for us to consider is not whether biopolitics is enabling or oppressing the subject, but rather that the power to subject is the power to create a category of abnormal objects. Where Canguilhem, Foucault's teacher, warned against the creation of the category of abnormal,[10] Foucault takes contemporary society as already committed to the task of defining the normal and the abnormal, and that commitment is perhaps no less obvious in the social arena than in the drive to define the object of disability.

On disability as object

I am of course playing into the hands of those who would create disability into a category unto itself. Numerous thinkers who identify as disabled have pointed out that disability should not be used as a categorical term. Blindness, or the inability to walk, or deafness, are all different conditions that lead to different perceptions and different modes of knowing. Thus, to lump together these different impairments of body under the category of disability must include the social and political decision to do so. Now the body is the body inscribed in society, for the only way to lump these different impairments together under a category of the same – that is to say under the category of disability – is to define how all of these different objects fail, not in a single bodily function, but in a societal function.

I want to pursue this a stage further. In fact, I would argue that the conditions for the possibility to lump these disparate conditions together within a category of 'disabled' in fact stems from medicine's own self-understanding. If physicians and physician-scientists are dealing with objects that do not function – physiologically or economically – then these disparate conditions can be categorized as one single category. If the proper functioning human being can see, hear, walk, talk – you name the feature – and then there are persons who are functionally limited in one of these categories, then that object – the eye, ear, legs/spinal cord, tongue – does not function properly. But the object is not the eye, the ear, the legs, the tongue. It is the body-machine that is afunctional, or malfunctioning because it does not function in society.

Thus, the necessary conditions to name an object as the disabled body – inability to walk, to hear, or to see – are only possible in a medicine that is the standard-bearer of Western metaphysics.[11] For these disparate features to be lumped together as a class of malfunctioning beings, one has to show that in some way they all lead to the same kind of malfunction. And since, as Foucault shows, medicine is a biopolitics concerned with the body-politic, the function moves away from mere biological function to social function, which for the West has become economic function.

Taking Foucault's (and others') point seriously, disability cannot exist except in a culture with its own political space, and its own history. Thus, it is possible – as many have pointed out – that, within the right set of social circumstances, a person with spina bifida and who is dependent on a wheelchair is not disabled. If the society has established ramps in all public buildings, he has the ability to thrive within that society and carry out his function; he is able to thrive. He is then able to function in society.

Yet, let's look at this a little more carefully. The language used is language of function. It remains a language that falls short of purpose, for purpose always transcends the immanent. A person functions in society if she can get around and can participate in the typical practices of that society. What this means is that because society has bought into Western scientific efficiency, into Western metaphysics, it will strive for functionality. Policies that allow people that lack some ability – an ability generally accepted as inherently good by dominant society – are not policies that promote a person's purpose within society, but allow for mere functionality. As long as a person can function in society, she can remain a member of society. The story goes that if society has to help them to function, then society should help those who are less fortunate. But if society can prevent from coming into being those who will not be able to function at this level – and in Western capitalist societies that means production – then it is legitimate to terminate a foetus who cannot function. And in fact, it is economic features that in practice lead people to the promotion of screening and prevention of disability through the termination of pregnancy.[12]

Disability creates such consternation for biomedicine and society because it is so confounding. As Martin Heidegger points out, it is confounding because an object is more present to the subject when the object is perceived to be broken.[13] And, it becomes merely an object because it is not inscribed within the relations of significance of the dominant society. In a society bent on production and where being able-bodied is necessary to production, the broken body is merely object, one that we are not yet ready to remove from before us, but one that nonetheless we would just as soon not see.

Yet it is not merely the non-functioning or dysfunctioning body that is perceived as abnormal. Biomedicine will strive to eradicate the body of the disabled, especially when they cannot be seen, when their costs can be prevented. The broken body becomes a useless object to medicine and society, and it becomes the negation of medicine's quest. Attempting to create usefulness by creating social policies that allow a modicum of functionality is understood as merely a stopgap measure, for those such as Harris. The disabled body must be eradicated, prevented prior to presenting itself before the medical gaze. The disabled body is an object that should not appear on the horizon of beings, because to see it face to face, changes the nature of its object status.

It is in this sense that modern medical science tends toward eugenics, preventing the coming, the appearing of all afunctional, non-functional, malfunctional, or dysfunctional objects. In short, Foucault is correct: the modern doctor with her understanding of subject–object, physician–patient, is not just a subject applying the Kantian transcendental and pure intuitions of space and time to an object in its raw facticity. The modern physician is a subject that constitutes its object against the unacknowledged categories of a political and economic space and as historically constituted. Function is always a function in society and economy, and if the goal of society is production of more capital for consumption, the object that cannot produce is useless in the realm of efficient causality. Functional economy is actually nihilistic.[14] Those that are not functional, those who cannot produce, are dispensable. Those who buy into society's functional goals will be consumed. From whence will come our salvation?

Part 2 The good or the ethical call of the other

Ethics and the objects of medicine

The answer to this question has often been a more ethical medicine. In its recent history, the bioethics movement began substantially from those outside of medicine who wished to maintain the 'patient as person'.[15] Thus, it was thought that the bioethics movement would have been in place to protect the interests of persons, or of the other. Yet, as pointed out over and over again, these critics were overtaken by the biomedical movement – bioethics becomes the handmaiden of biomedicine. What results is the proceduralist ethics that (rather thinly) states: 'Let's be sure to have transparency, the proper oversight or surveillance procedures, and we will get good ethical medicine or medical research'. As many others have pointed out, a common morality for all is very superficial and thins out the richness of religious traditions,[16] which may be precisely the point for the proceduralists.

In a way, the proceduralist ethic works hand-in-glove with biomedicine. If biomedicine is on the quest to assure function within the realm of efficient causality, then it becomes necessary for those who would manage the functional processes to be managed by those who would assure proper function of the medicine within the social and ethical realm. And the result is that a proceduralist ethics paves the way for biomedicine to promote its functional goals within society, subsequently attenuating the fears of society.

The bioethicist comes to have a two-faced role. First, she has the role of protecting research subjects or patients – or more precisely, the manipulated objects of medicine – who are subjected to the agency of the researchers and doctors; and second she has the role of assuring the public that doctors and researchers are behaving themselves. There have been several critiques lately that criticize the complicity of bioethics with medicine's ideology.[17] A proceduralist ethics is complicit with medicine's goals.[18] Biomedicine, for the most part, has realized the necessity of bioethics and, in many medical schools, departments have been set up, bringing the regulators within the medical fold. Thus, it has become those in departments of medical ethics, or bioethics, who have been vociferously attacking the position of the disability rights groups.[19]

As pointed out before, there are those who cannot bring themselves fully to dispense with disabled bodies, for in the encounter one experiences the truly other in their brokenness. Martin Heidegger has been tremendously important to metaphysics. Western science has argued that objects are just there for investigation. Objects and questions about objects are always already there within a field already rife with meaning and are therefore not just objects. Heidegger makes an interesting point. It is the very fact that an object appears as broken that draws our attention to it.[20] That is to say, the (mal)functioning body becomes conspicuous for its inability to carry out its function in economy.

But the otherness of the abnormal foetus can be bracketed with the machines that have extended to gaze into the womb. With a test such as a blood-test or even a sonogram, the other is made more present in one way and yet more distant in another way. In their brokenness, the disabled show themselves phenomenologically as radically other, but with the extension of the gaze through technology, the other is prevented from appearing to the doctor. If those with broken bodies appear, then there is resistance to the gaze. But technology is so politically useful that there can be little resistance. Because the disabled fail in efficiency and effectiveness, that is to say, in that they fail to meet the demands of the dominant political and economic society, they fail as objects and become truly other for dominant society. So while the scientific gaze creates the disabled body because it does not fit into the categories of normal societal functioning, something of the phenomenality of the broken body convicts those who would annihilate the disabled body that cannot be seen. They cannot annihilate these others who appear in their brokenness. The subject is left to converse, to come out of its dominating gaze, to engage those with broken bodies. But this is not the case for those that appear by the extension of the gaze with technology.

It is here that the work of Emmanuel Levinas becomes important. For Levinas, it is the other that calls the subject – the physician – out of her constituting gaze, out of the immanent realm of cause and effect. In a way then, it is the broken body that might be the salvation of medicine. This other resists and defies objectification.

The other, or otherwise than being[21]

Emmanuel Levinas both embraces and rejects Heidegger. He notoriously gives praise for Heidegger's insight, and criticizes his personal and professional failures. Yet, despite Heidegger's failures, he cannot dispense with him.[22] In a sense, Levinas is attempting a different kind of transcendental philosophy. He is seeking a more primordial and/or primitive condition than the condition of being-there, of *Da-sein*. He seeks to articulate the experience of what he describes as the 'call to holiness preceding the concern for existing'.[23] He seeks the 'no-place prior to the there of being-there'.[24] In a way, being-toward-death, the death of the other owned by *Dasein*, became, for Heidegger, a spectacle, for the preparation of authentic existence. Levinas castigates Heidegger for not seeing the call to responsibility in the face of the other. He condemns Heidegger because the call of Being is not a call to respond to the death of the other. In this way, Levinas escapes the primacy of ontology, of beings, of objects.

In *Totality and Infinity*, Levinas speaks of the absolute transcendence of the other. This transcendence is articulated in the vulnerability of the face. But the whole body is face, but not as some thing perceived. When one experiences – experiences, not sees – the face, one is called to respond, to speak. Thus one is called out of oneself in speaking to the other. But the response is not just a verbal response, it is moral response. The call is thus a call to moral responsibility. The other escapes the categorizing gaze of the subject. The other calls the totalizing subject to response and to responsibility. The face announces both closeness and necessary distance. The call to justice is the religious call that binds – *religare* – one to the other but without totality. It is not cognizance, nor is it ontology, but prior to both and the necessary condition for the possibility of either knowledge or being.[25]

In *Otherwise than Being* (1969), the 'I' is placed infinitely subject to the other. If others do not do what they are supposed to do, then I am guilty as well. The 'I' becomes responsible for the activity of the other. And it is absolute, for I am responsible for the other and guilty for the inadequacy of the response, and guilty of the actions of the other as well. Levinas fondly quotes Dostoevsky: 'We are all guilty of all and for all men before all, and I more than the others'.[26] In the ethical relationship, the 'I' is truly subject. The 'I' does not subject as in the relationship of knowing, or doing science or medicine, for that is the quest for the 'real' and the 'true'. Rather, the 'I' becomes subject to the other, and that is a hungering after the good.

Levinas's greatest contribution to phenomenology, however, is not his ethics, but the Copernican revolution he brings about through a reversal of the intention. While with Husserl's[27] and Heidegger's phenomenology,[28] the intention is directed outward, with Levinas we find that the other's intention toward its object 'me' takes precedence. The other constitutes me.[29]

With regard to disability, there is a sense of the face. As pointed out, with disability, modern medicine sees an object – an object subjected to the constituting gaze of biomedicine. When gazing into the womb with all the objectifying tools of medicine – whether sonograms, prenatal diagnostic tests and screening – the foetus cannot call for response. And thus the turn to eradication is easy, for without face, it is easy to perceive the biological problem and to reduce it to the status of object. The foetus does not speak, does not call. Disability – medicine's own progeny – disability, seen in the womb, remains merely object.

The death of the one and other

Still, for all of the eloquence of Levinas, there are some problems that we do not escape. In what way does the face appear if not in discourse or in experience? In what way are we responsible for all? In what way does the other call, when the other cannot engage? In what sense does mental impairment appear as other when this other does not have capacity to converse? Before turning specifically to disability, or better, impairment, I want to look a little more specifically at Levinas.

Levinas is deeply Kantian in many regards: a criticism that he would no doubt find atrocious. First, he is attempting to move away from ontology arguing for the primacy of metaphysics. By this, he means that reason is left to speculate about beings in the same way that pure reason is left to speculate about the noumena for Kant. One is left only with phenomenality in the end and the object is never really exhausted by the intellect. Thus, the more primal relation is the ethical relation, just as for Kant the more proper use of reason is practical reason, that is to say the proper field for reason is the moral. Just as for Kant proper freedom is freedom in line with the dictates of duty, in a sense, the proper origin of freedom in Levinas is when it is in response to the call, the demand of the other. While Levinas is transcending Kant, at the same time he does not escape some Kantian pitfalls.

Again, in Kantian morality, our moral activities are never truly moral unless they are acontextual, unless they derive solely from the categorical imperative. Like the categorical imperative, the face of the other, which carries the command, 'Thou shalt not kill', but also in its vulnerability 'invites us to an act of violence', is also without context. 'The face is signification, and signification without context.'[30] The response is a response without hope of return. An act of responsibility to the other cannot be good unless I receive no benefit from it. Just as for Kant, an action that has a return may not be a moral action unless the response stems from the imperative, Levinas longs for the possibility of no return. Yet, the hope of no return seems grounded on the need to outdo the other, and if not to outdo, then to mask what returns.

The call to responsibility is so demanding that sin is necessary, and failure inevitable. It is not just the I–Thou relationship. Levinas states:

> In fact, such a society consists of two people, I and thou. We are among others. Third parties are excluded. The third man disturbs this intimacy essentially: my wrong with regard to you, which I can recognize entirely in terms of my intentions, is objectively falsified through your relations with

him, which remain secret to me, since I, in turn, am excluded from the unique privilege of your intimacy. If I recognize my wrongs with regard to you, I may be wronging the third one through my repentance itself.[31]

The 'I' and the 'Thou', the relationship is not the 'beginning of society, but its negation'.[32] According to Levinas, this duo of 'I' and 'Thou', is the 'closed society'.[33]

While the call is the call of infinity, there cannot be a single relationship between the 'I' and an other, for in excluding others one sins. Yet, with the call of infinity are we doomed to failure, to sinfulness? Is it not this particular other to be sacrificed, to be negated in the midst of all the others? Certainly, a Levinasian ethics requires forgiveness. Yet even forgiveness is the transactional opposite of vengeance for Levinas – both compensating for what cannot be compensated, and thus failing in the face of the call of the other.

This ignoring of the particular relation of the 'I' and the other is, in a way, a usurping of the limits of particularity and responsibility. In itself, the demand of others calls for the death of the one who is with me – that is to say, for the other to negate itself in the midst of all the others. Love must remain anonymous, generic and, most certainly, never particular. The 'I' who loves must not love particularly. There can be no true responsibility in any ethically important way, for the demands of the particular relation are lost in the overwhelming infinity of all the others. Is it not just as sinful to will the infinite, a drive to the infinite at the expense of the particular, a drive to the universal imperative?

And here is where Levinasian ethics fails. There seems to be no place for love, for grace, for givenness beyond measure, only the demands of disinterestedness of justice.[34] For Levinas, phenomenologically speaking, the appearing of the other is the appearing of an empty form, leading to negation and nihilism.[35] It is the vulnerability of the other, the radical possibility of his death, for which I am responsible, that allows her to appear for me as other than object. She overwhelms totality with her infinite but unanswerable call. She overwhelms the same; she overwhelms the will to know and categorize to say that she is same as this general category. That is to say, she overwhelms me. I must die. Or as John Milbank puts it, the self is eroded in a meaningless self-sacrifice, in the death of the self.[36]

Death, thus, is the necessary condition for the ethical. But it is not only the death of the one called, but also the death of this other. And moreover, I am doomed to kill all the others who likewise lay claim to my response. She is, thus, doomed to death, to disappear. She is doomed to annihilation. The condition for the possibility of the appearing of the other is the disappearing of this particular other. Perhaps Levinas has escaped the problems inherent in ontology in order to do metaphysics through his ethics, but he creates an impossibility for an ethics of particularity. It collapses under its own weight. In short, as pointed out by Jean-Luc Marion, for all of the brilliance of his phenomenological mentor, Levinasian ethics fails because the ethical horizon fails to individuate the other and collapses into a different kind of, but nonetheless Kantian, formalism.[37] The neighbour remains far away, distant.

Finally, in the real world of particular relationships, Levinas's ethics seems silent before the face of those who cannot engage in discourse. Those who are hidden away by society and from society, those who cannot engage totality at all, and are left to be totalized by utilitarian, efficient and proceduralist ethics and technological gazes. First, the other is not encountered except in the obscurity of the sonogram, the indeterminacy of bioassays, the discourses of the human sciences, or the structural rigidity of the intellectual assessment. These instruments of the gaze cannot ever let the impaired body be, for they exist and are created to highlight the impairment, to bring it into relief against the horizon of beings. Those whose impairments are more profound, or who cannot show themselves from the womb, are silent, precisely because the call cannot be heard over the chatter created by the gaze or the discourses that construct the objects of the gaze. For Heidegger, the work, the already defined quest – and resultant questions – comes to let things and others appear for the subject. For Levinas, the other comes to shape the work in which the other is engaged, but, as I have pointed out, only to annihilate the possibility of any particular call or any particular response: that is to say, only to annihilate any particular relationship. For these silent ones and those who care for them, there is an all together more primordial feature at play than a mere object that appears against the horizon of being, and something far more profound than the call of an articulate other against the horizon of ethics. Ethics fails, and will fail, to be the saviour of society generally, and of medicine particularly. For the technological extension of the gaze fails to let be what is given particularly in experience of the call. Phenomenologically speaking, we fail to attend to the experience with the horizon of givenness; theologically, we fail to notice the gratuity of what is revealed in experience.

Part 3 The beautiful, or the return to aesthetics

Marion's phenomenology

Rather than thinking in terms of subject–object, we must pay attention to the givenness of experience. What might this entail? First, all theoretical pursuits must be bracketed – a task in which I am not sure that medicine can succeed.[38] This bracketing is almost impossible because medicine has increasingly extended the gaze through technological developments that compel both technological categorization, as well as discursive categorization within a society bent on production. Medicine is steeped in these modes of seeing and looking. Second, we should take notice of the possibility of the dependence of the subject upon what is given, that even intention (in Husserl's sense) is given. As stated, Levinas's contribution to phenomenology was the reversal in directionality of intention. This task too will be difficult for medicine. The ability to look, the ability to categorize before the appearance of the phenomenon of the body of the other is the ability to know the object before it is even a phenomenal object. Can it ever really appear for medicine?

In short, we are asking a lot of medicine. We are asking if the constituting gaze of the doctor might become the constituted subject-physician. Medicine will first have to admit, and it is odd to say it, that the physician does not appear until the other gives her (the doctor) her constitution, by calling her into being, *qua* physician.

In this sense, we must learn to speak of the revelation of the person – the person as icon. Here I draw on the distinction made by Marion in God without being between idol and icon.[39] Idol is what is constituted by me, the subject. It is the application of categorical knowledge to the object. The idol emerges as an object constituted by what Foucault calls the historical a priori. I constitute the idol because I subject the given phenomenon to my categories of analysis. In the case of the disabled body, the idol emerges as an object only because it has been subjected to the categories created by a scientific medicine complicit with a society bent on material production. It is a worshipping of one's own creation. It is the constituting act of the subject. There is a terrible caveat to the creation of the idol; it is not 'real', for the phenomenon has never been let to be. And this terrible fact, this figment of our imagination, this idealized god becomes the measure against which the broken body is compared. When the broken body cannot meet the standards of this gaze, when the broken body does not live up to the idealized god, it becomes the disabled body. And the terribleness is that the broken body can be sacrificed to the idol. It can be annihilated.

An icon, on the other hand, is that which comes to constitute us. It must be left to be, to become what it will be. In the icon, I do not see God. God sees me. In the icon, the directionality of intention (Husserl) or the directionality of the gaze (Foucault) undergoes the Levinasian reversal. In the icon, I become, not the subject, but the subjected. Through the icon, I am called out of my objectifying stupor. In this sense, the broken body could be the icon for medicine. But why is it that the broken body does not call the doctor out of her objectifying stupor and into becoming what she is? The subject must change from she-who-subjects to she-who-receives. In this sense, it is doctor who must be constituted by the gratuity of what has been given in this particular other. What has been given is the doctor's very being, for the other calls the doctor to become who she will be, and this call comes through the icon of the broken body. Against the horizon of givenness, it is this particular other who calls us into being persons.

However, there is always the risk of Abraham. It is not easy to be called out of yourself. It is not easy to be constituted by an other, foreign, yet so close. It may demand all from us. But the gratuity of the experience, even in the negation of that which we had hoped for – the brilliant musician child who turns out to fall into the Down's category, the brilliant physicist who turns out to fall into the category of intellectual impairment – remains palpable if not present, and perhaps precisely because not present. What we had hoped for becomes our idealized god, and if the broken body cannot meet the categorical demands, it can be annihilated. We are left either to measure the experience of the particular other against the idealized god – the brilliant musician or physicist – the idol of our own construction. Or, we can be called into becoming what the other calls us to be, needs us to be.

And what of the unborn child hidden from view? How is she icon? When she moves within her mother's womb, the mother occasionally shrieks in horror at the ownership the child within her has over her. If I may be so bold – as a man, to presume such things is often difficult – if she sees with the extended gaze of technology the lack of all her desires for the child, she sees the absence of her idol. It can be annihilated. But most often – and fortunately for us, much more often – when the child moves for the first time she becomes mother. Everything is different now. She is constituted by the other in responding to this call. She is created as mother of this child by the call of this child. Not only is the child given, she stirs the mother's desires, her love. This stirring of desire is dependent upon the givenness of the experience.[40] What is given in the experience is not only the phenomena and the response, but a desire for the good of this particular other, and this one thing more must come first.

For medicine, then, there is a difference between seeing a disabled body and the appearance of a broken body; that is to say, the difference is between seeing a body that does not fit into our categories, and a body that calls us – all of us, as carers – to become what that body of the other needs. That the technological gaze, indeed that medical education itself, occludes and prevents the phenomenal appearing of the broken body, is only indicative that medicine is indeed the standard-bearer of Western metaphysics, as Krakauer has claimed.[41]

The contribution of phenomenology (Husserl) has been to ask the sciences to bracket the theoretical perspectives in order to attend to the phenomenon, to the experience. Heidegger has asked us to attend to the call of Being, rather than the beings. Levinas's contribution has been to reverse the intentionality, to focus on the call of the other. Marion's contribution has been to show that all of these are given in the horizon of givenness. And in the gratuity of the given there is one more thing given; always there is one thing more. Even the desire for the good of the particular other is given in the experience. As Marion states, 'only love opens up knowledge of the other as such'.[42] The question is, can medicine even hear or understand the necessity of desire, of love, since it continues to focus on that which appears as the object of its gaze.

I have argued that we have prevented the broken body from appearing, because we have looked for the disabled body. I have argued the disabled body is an object that is known precisely because of the idealized body. It does not live up to the image of the false god of medicine. In short, the icon for all carers – the broken body – is what calls us into becoming what the particular other needs. Yet, medicine has both technologically and discursively extended its gaze, and in so doing it has numbed its senses. It has become anaesthetized, literally an-aesthetized, without sense, outside of the senses. In a way then, the call is for medicine once again to return to its senses,[43] to return to the senses, once again to experience aesthetically, for to attune to what is given in the senses is to return to aesthetics. Once again to attune to what is given is to let one's desires be stirred by the gratuity of what is given.

I began by stating that no medical student chooses the career of medicine because she wishes to objectify a human being. No student ever starts out by stating that she wishes to subject her patients to the categorical knowledge of field of medicine. Inevitably in every medical school application the student lays out what she believes to be a call. What gives the call? It is – often enough to hold out hope – the love of an other that draws her into medicine. It is the love of a parent, a grandparent, sibling or friend whose suffering was witnessed, whose suffering calls her to become physician, whose death calls her into being. There is a desire to relieve suffering. While this call might be interpreted as the need for technological intervention, there is still the givenness of the absence of the loved one – a love without being, as Marion puts it.[44] The givenness that suffering carries with it calls for response. And the prospective student responds to what she has been given – an iconic gift. And like all gifts, they hold sway over us. If we can return to that original call, that is to say, if medicine can return to its senses, there may yet be hope for medicine.

Theology, and the education of desire

I have argued that the gaze comes to order and shape what is seen by the doctor. And I think that this is most certainly the case with disability; the body becomes object precisely because the disabled body does not function efficiently for societal production. To this point, I have offered phenomenology as an alternative to the objectifying gaze of medicine. Yet, within phenomenology, this call to bracket theory, organizing discourses, and categories of normalization comes as almost a romantic appeal to some original, pre-theoretical insight, or some original experience that could reorient medicine. At one level, there is the sense that it is possible to redefine medicine through better philosophy, one that allows us to rid ourselves of the concretions of cultural and societal theorizing, categorizing, and normalizing. But this call for original experience will not return us to the golden age of phenomenology, a pristine philosophical point where medicine might recapture its nobility. Instead, just as we learn to theorize, to think in certain ways, we can become practised at letting things be, letting the other define our call, but not in some recast Kantian formalism; we can become practised at letting the broken body call us out of our objectifying stupor to meet its needs.

Even while medicine is not a theory-generating activity, it nonetheless uses theories. Doctors come to patients with certain theoretical and practical stances, informed by various kinds of knowledge about the body, bodily function, societal expectations, sociological information, and so forth. But medicine is a practice and one becomes practised in it by doing it. Thus, one learns to apply the gaze, to categorize, to thematize and to theorize. One is taught, is socialized into application of the gaze. And it is here, that theology is indispensable, for what is given is not just practice at letting be, but also the disposition of receiving. Theology is not a set of systematically consistent doctrines, but a set of practices in prayer, practices that teach and orient the practitioners to let be and to receive the call of the other. Divine liturgy is the practice of letting God constitute oneself. In liturgy, the subject is constituted by God.

Marion's work has been criticized by those in phenomenology who see theology as bringing its own sets of theories/doctrines to bear on experience, the content of which is revealed, and also as going beyond the methodology of phenomenology.[45] I have accepted Derek Morrow's reading on this, that Marion in fact survives the critique and that his methodology is faithful to phenomenology.[46] Still, there is something more that theology gives, and that is that in its practices – particularly in the practice of liturgy – there is a stance toward givenness, towards the gratuity of excess that can be learned. Indeed, for the entire Neoplatonist tradition, particularly the Augustinian Christian tradition, there is the sense that one must be rightly disposed in order to receive rightly the gift.[47] In a sense, one can be, indeed must be, educated toward givenness through the practice of liturgy. One can be habituated to receive what is given. It is here that Marion's givenness fails, for one can rightly receive and respond to that which is given, but it is possible that one can receive wrongly. Thus, we must practise to be rightly disposed to the other.

It is not, however, to the rationality of liturgy or to its structures, or to the constitutive act of divine liturgy that I turn, for these are but the preparation for the appearance. In divine liturgy we are constituted so that we can receive he who is truly other, calling us out of ourselves, not in the annihilation of ourselves, but in the giving of ourselves to ourselves. Likewise, this gift is the gift of hope, necessary to overcome the nihilism (the death of the one and the other) in modern self-giving ethics. In addition, this other is not a formal Kantian other, but is fully embodied, fully present. And that which constitutes the subject, calling her out of herself into what she is to be, that which stirs her desires to be what the other would have her to be, is also that which appears: the broken body of Christ. Coincident in the broken body is the reconciliation of science, ethics and aesthetics – the true, the good and the beautiful.[48] In divine liturgy, we are given our desires, the mandate for the other, through the giving of what can now appear for us, that which is given for us – the broken body.

Contemporary 'secular ethics' tends to understand religious prohibitions against interventions such as abortion or euthanasia as irrational, rule-based ethics, adherent to some deontological category. However, it should be clear by now that religious prohibitions against abortion and killing in all of its forms – that is to say, the annihilation of those who do not meet the measure of the idealized god, the idol of modernity – are derivative from the gift of divine liturgy. Instead, they do not understand that religious people are formed by the practices of openness that divine liturgy affords, openness to the gratuity of all challenges, of all the calls of each of the others.

Divine liturgy challenges modern subjects; the broken body comes to constitute the subject, to subject the worshipper to the divine seeing, to the divine gaze. Our desires are educated, prepared, not in the sense that we are habituated into loving certain kinds of things. Instead, our desires are educated, in that in divine liturgy our desire to let God call us into becoming what God would have us be, is itself first a formation of desire, an ever-changing process of refinement. In divine liturgy, we learn our dependency upon gracious gifts as creatures.

If there is one thing that disability rights scholars and activists have shown medicine, it is that we are not autonomous agents but instead dependent creatures. As Christopher Newell, one such scholar, states: '[I]t is perhaps best to think of us humans primarily as dependent animals, rather than as rational autonomous wills asserting sovereignty over our bodies. We who live with impairment know this better than most.'[49]

Through the practices of divine liturgy, we are constituted as dependent beings called to be for the other. In the here and now of divine liturgy, which is also the 'all times and all places' of divine liturgy, all is given, all is reconciled. And ethics and science, and the rightly oriented desire for the good of the other, are once again possible.

Notes

1. I am particularly grateful to my friends and colleagues Alan Bleakley, Philipp Rosemann and Derek Morrow for their comments on and discussions of the topics of science, ethics and aesthetics addressed in this chapter.
2. Witness the protest staged by Not Dead Yet on 13 July 2006 at the American Society for Bioethics and Humanities' Bioethics and Politics conference at the Alden March Bioethics Institute in Albany, New York.
3. E. Krakauer, 'Prescriptions: Autonomy, Humanism, and the Purpose of Health Technology', *Theoretical Medicine and Bioethics*, 19 (1998), pp. 522–45, 535.
4. By modern, I mean the natural sciences as they began to emerge out of early modern philosophies. These philosophies seemed adequate until recent findings in relativity and quantum theories began to take hold. I am arguing that medicine still holds to these early modern theories. E.A. Burtt, *The Metaphysical Foundations of Modern Physical Science* (Amherst, NY: Humanity Books, 1999). (First published in 1924.)
5. Krakauer, 'Prescriptions', p. 535.
6. E. Levinas, *Otherwise than Being, or Beyond Essence* (trans. Alphonso Lingis; Pittsburgh, PA: Duquesne University Press, 1998). (Originally published in 1974, under the title of *Autrement qu'être*.)
7. Krakauer, 'Prescriptions'.
8. M. Foucault, 'La Crisis de la medicina o la crisis de la antimedicina', *Educación Médica y Salud*, 10.2 (1976), p. 152.
9. For this point, I am thankful to Philipp Rosemann and Peter Candler for previous and ongoing discussions.
10. G. Canguilhem, *The Normal and the Pathological* (New York: Zone Books, 1989). (Originally published as *Le Normal et le pathologique* [Paris: 1966].)
11. Krakauer, 'Prescriptions'.
12. W.J. Gagen and J.P. Bishop, 'Ethics, Justification, and the Prevention of Spina Bifida', unpublished paper, accepted for publication in *Journal of Medical Ethics*.

13. M. Heidegger, *Being and Time* (trans. Joan Stambaugh; New York: State University of New York Press, 1996), pp. 67/72–71/76. (Originally published in 1927, as *Sein und Zeit*.)
14. D.S. Long, *Divine Economy: Theology and the Market* (New York: Routledge, 2000).
15. P. Ramsey, *The Patient as Person* (New Haven, CT: Yale University Press, 1970).
16. H.T. Engelhardt, *The Foundations of Bioethics* (Oxford: Oxford University Press, 1996); H.T. Engelhardt, *The Foundations of Christian Bioethics* (Lisse: Swets & Zeitlinger, 2000).
17. D. Callahan, *What Price Better Health: Hazards of the Research Imperative* (Berkeley, CA: University of California Press, 2003); W.J. Smith, *Culture of Death: The Assault on Medical Ethics in America* (San Francisco, CA: Encounter Books, 2000).
18. A.L. Hall, 'Whose Progress? The Language of Public Health', *Journal of Medicine and Philosophy*, 31.3 (2006), pp. 285–304.
19. J. Harris, 'Is There a Coherent Social Conception of Disability?', *Bioethics*, 9 (July, 1995), pp. 3–4; D.W. Brock, 'The Non-identity Problem and Genetic Harms: The Case of Wrongful Handicaps', *Bioethics*, 9 (July 1995), pp. 269–75.
20. Heidegger, *Being and Time*, 67/71.
21. I refer, of course, to Emmanuel Levinas's work, *Otherwise than Being* (see n. 6.)
22. Levinas states: 'Despite all the horror that eventually came to be associated with Heidegger's name – and which will never be dissipated – nothing has been able to destroy in my mind the conviction that the *Sein und Zeit* of 1927 cannot be annulled . . .' E. Levinas, 'Dying for . . .', in *Entre nous: Thinking-of-the-other* (trans. Michael B. Smith and Barbara Harshaw; New York: Columbia University Press, 1998, pp. 208, 207–17. Speech delivered in 1987 at the Collège International de Philosophie in Paris. First published in 1998 under the title *Heidegger: Questions ouvertes*.)
23. Levinas, 'Dying for . . .', p. 216.
24. Levinas, 'Dying for . . .', p. 216.
25. E. Levinas, *Totality and Infinity: An Essay in Exteriority* (trans. Alphonso Lingis; Pittsburgh, PA: Duquesne University Press, 1969). (Originally published in 1961 under the title *Totalité et infini*.)
26. F. Dostoevsky, *The Brothers Karamazov* (trans. Constance Garnett; New York: American Library, 1957), p. 264; E. Levinas, *Ethics and Infinity: Conversations with Philippe Nemo* (trans. Richard Cohen; Pittsburgh: Duquesne University Press, 1985). (Originally published in 1982 under the title, *Ethique et infini*.)
27. E. Husserl, *Cartesian Meditations: An Introduction to Phenomenology* (trans. Dorion Cairns; Dordrecht: Kluwer Academic Publishers, 1950). Originally published in 1931 under the title *Méditations Cartèsiennes*; E. Husserl, *The Crisis of European Sciences and Transcendental Phenomenology* (trans. David Carr; Evanston, IL: Northwestern University

Press, 1970). (Originally published in 1952, under the title *Die Krisis der europaischen Wissenschaften unde die transzendentale Philosophie*.)
28. Heidegger, *Being and Time*.
29. J.L. Marion, *Being Given: Toward a phenomenology of givenness* (trans. J.L. Kosky; Stanford, CA: Stanford University Press, 2002), pp. 266–69. (Originally published in 1997 under the title *Etant Donné: Essai d'une phénoménology de la donation*.)
30. Levinas, *Ethics and Infinity*.
31. E. Levinas, 'The I and the Totality', in *Entre Nous: Thinking-of-the-other* (trans. Michael B. Smith and Barbara Harshaw; New York: Columbia University Press, 1998), p. 19. (Originally published in *Revue de metaphysique et de morale*, 59.4 [October–December 1954], pp. 353–73.)
32. Levinas, 'The I and the Totality', p. 20.
33. Levinas, 'The I and the Totality', p. 21.
34. Levinas, 'The I and the Totality', pp. 19–20; Levinas, *Ethics and Infinity*, pp. 99–100.
35. Marion, *Being Given*, pp. 266–69; J.L. Marion, *Prolegomena to Charity* (trans. Stephen Lewis; New York: Fordham University Press, 2002), p. 165. (Originally published in 1986 under the title *Prolégomènes à la charité*.)
36. J. Milbank, *Being Reconciled: Ontology and Pardon* (London: Routledge, 2003), p. 145.
37. D. Morrow, 'The Love "Without Being" That Opens (to) Distance [Part 2]: From the Icon of Distance to the Distance of the Icon in Marion's Phenomenology of Love', *Heythrop Journal*, 46 (2005), pp. 493–511.
38. Husserl, *Cartesian Meditations*; Husserl, *The Crisis of European Sciences*.
39. J.L. Marion, *God without Being: Hors-text* (trans. Thomas A. Carlson; Chicago, IL: University of Chicago Press, 1991). (Originally published in 1982 under the title *Dieu sans l'être: Hors-texte*.)
40. For an excellent summary of Marion's work and an introduction to Marion's *Le Phenomene Erotique: Six meditations* (Paris: Grasset, 2003), see D. Morrow, 'The Love "Without Being" That Opens (to) Distance [Part 1]: Exploring the Givenness of the Erotic Phenomenon with Jean-Luc Marion', *Heythrop Journal*, 46 (2005), pp. 281–98; Morrow, 'The Love "Without Being"', (Part 2). See also, Marion, *Being Given*.
41. Krakauer, 'Prescriptions'.
42. Marion, *Prolegomena to Charity*, p. 165.
43. A. Bleakley, R. Marshall and R. Brömer, 'Toward an Aesthetic Medicine: Developing a Core Medical Humanities Undergraduate Curriculum', *Journal of Medical Humanities* 27 (November/December 2006), pp. 197–213.
44. Morrow, 'The Love "Without Being"', 2005a and 2005b.
45. D. Janicaud, *et al.*, *Phenomenology and the Theological Turn: The French Debate* (trans. B.G. Prusak; New York: Fordham University Press, 2000).

46. Morrow, 'The Love "Without Being"' (Parts 1 and 2).
47. J. Milbank, 'Only Theology Overcomes Metaphysics', *New Blackfriars*, 76.895 (1995), p. 332.
48. I am of course dependent on John Milbank here and his thesis in Milbank, *Being Reconciled*.
49. C. Newell, 'Disability, Bioethics, and Rejected Knowledge', *Journal of Medicine and Philosophy*, 31.3 (2006), pp. 269–83.

Conclusion: Fragility and Grace; Theology and Disability

Robert Song

All of the contributors to this volume have reflected in different ways on a single set of themes, linking the experience of disability with advances in genetic science and medicine and with questions about eugenics. Whether reflecting on firsthand encounter explicitly or tacitly, or engaging with the issues through the disciplines of biological science, clinical medicine, history, philosophy or theology, each of the chapters forms part of what is cumulatively an extraordinarily rich and densely woven fabric of thought and personal experience. In this concluding chapter, I will draw out and comment on just a few of the book's threads, in full awareness that it contains a far greater wealth of material than can be discussed here.

Genetic science, disability and prenatal testing

Twentieth-century breakthroughs in human genetics hold out enormous promise for our understanding of human beings and for the future of medicine in particular. The reception of Mendel's account of the laws of inheritance, the discovery of the structure of DNA, the development of various technologies of genetic manipulation and the sequencing of various genomes, have just been some of the highlights of the burgeoning science of molecular biology. Following on from these, the identification of several chromosomal and genetic disorders, together with the possibilities for genetic screening, gene-therapy, personalized medicine, cloning and stem-cell research, and all the associated epidemiological and statistical techniques, are beginning to alter medicine permanently.

This story, accounts of which are found in the chapters by Walter Doerfler and Blair Smith, is one of the sustaining narratives of modern Western culture. It reassures us who inhabit that culture that there is hope for deliverance from some of the things we most fear: sickness, disease and the suffering which accompanies them. It sanctions the power and public standing of those institutions, the great organs of research into molecular biology and medical science, which promise to wrest that fire from the gods and bring it down to mortals. It elevates to immortal memory the heroic pioneers who by their courage, persistence and ingenuity have laid bare hidden secrets and earned for themselves the title of benefactors of humanity. Despite the temporary hold-ups and perils that will no doubt be encountered along the way, the background narrative of modern scientific medicine offers the consoling prospect of a future increasingly free from suffering and the imperfections of body and mind that haunt human existence.

Of course the story can be told unpresumptuously, without raising expectations recklessly or holding out implicit promises of cures for currently incurable conditions. Smith, for example, makes clear in his contribution that the elimination of disability *tout court* is not an aim of genetic science, nor even a possibility without eliminating the human species, whereas the more modest goal of alleviating suffering is both intelligible and achievable. The measured prose of government papers and policy statements on health-care can similarly be guaranteed to eschew the more feverish speculations about the future – even if only for reasons of coyness about available financial resources. It is in principle possible, in other words, to give an account of our societal commitment to health-care which places the practice of medicine firmly within an understanding of the care which vulnerable, mortal human beings owe to each other in their suffering. The provision of health-care need not be portrayed in a messianic light: it can be subordinated to a recognition that the most valuable things in life will never be achieved by medical science, however marvellous this might become.

Yet, although the restriction of medicine to such a subordinate role remains a standing possibility, one may still wonder whether in contemporary Western societies even the commonplace language of health policy – 'improving the nation's health', 'reducing the number of deaths from disease', and the like – in fact surreptitiously contributes to a different mentality. After all, cost-conscious governments still have an interest in conveying a sense of rising health outcomes. And a citizenry who have been taught to see themselves as consumers of health-care have become only too ready to view good health in themselves or their children as more nearly a right to be demanded rather than a gift to be cherished. Medicine and health policy are not delivered in a vacuum of expectations, but are inextricably bound up with the desires which structure dominant patterns of social and cultural power and are in turn structured by them. Such hopes and fears in the area of health inevitably cluster around suffering, disability and death, and the chances of warding them off. And in relation to such matters, the prospects for humility of human aspiration, perhaps especially in a society increasingly orientated to the rewards to be had in this life, are not strong.

This creates an unavoidable tension between the insatiability of consumer demands on the health-care system and the limits of that system's resources, a tension that is worked out on the backs of the clinical staff and health-service managers whose daily work is to steward these conflicting demands. Perhaps nowhere is this clearer – or, as we shall see, more ironic – than in relation to public attitudes and health policy in relation to disability. In general terms, most Western countries now offer to disabled people as good and imaginative health-care, social provision and rights-based recognition as has ever obtained in history. Those resources are not unlimited, as is only too familiar to service-providers and service-users: the lack of adequate work, failures of local transport, sometimes punitive welfare benefits, and problems in access to health services, are all typical examples. But in the context of health-care there is a rather different way in which these constraints are also intimated, namely in the practice of prenatal testing. The ever closer surveillance of pregnant mothers through a relentlessly advancing armoury of tests and procedures has become a routine and expected part of the experience of pregnancy. To be sure, government spokespeople have been known to deny that these are instituted on budgetary grounds, but it beggars belief that such gruelling rounds of investigations, in the vast majority of cases of which the only 'therapy' – and the usual outcome – is abortion, would be undertaken if there were not the expectation of significant financial benefit.

Apparent limits on resources are far from being the only reason for the regime of prenatal diagnostics. Even though their son Adam was diagnosed as having Down's syndrome only after birth, and therefore long after they could be advised to have an abortion, Brian Brock and Stephanie Brock describe in their chapter how they still faced cases of Adam being tested without their informed consent, undergoing experiences of confrontation with doctors and even the outrage of medical staff. Here, the Brocks explain, the issue underlying medical anxiety was a different perception of the nature of responsible parenting, according to which it is the job of conscientious parents to embrace genetic testing. It was not the inhumaneness of individual medical practitioners that was in question: indeed it was their very humaneness that led them to be so convinced about the neutrality of genetic information and therefore the unproblematic nature of having more of it.

It is not of course hard to speculate about the reasons why responsible parenting might be thought to require genetic testing, just as it is not hard to be sceptical about the mantra frequently repeated to the Brocks that 'testing is not a judgement on disability'. The everyday experiences faced by disabled people, ranging from stares and children talking over their shoulders as they pass by to the well-meant but all-too-frequent comment, 'What's wrong with you?', as Christopher Newell relates, as well as the experiences of ignorance, prejudice and discrimination in schooling and health-care, housing and employment, all play their part in shaping the general perception, mirrored in both public and professional responses, that disability is invariably a tragedy. It is scarcely surprising that despite successes in removing the more obvious architectural and physical barriers to social participation, the disability rights movement continues to insist on distinguishing physical or mental impairment from socially imposed disability.[1]

But the cultural context within which disability is constructed extends beyond the facts of prejudice and social and economic discrimination to the practice of scientific research itself. Both Newell in his contribution, and Brian Brock, Walter Doerfler and Hans Ulrich, in the report of their collaborative conversation, show how genetic research is being used in the creation and reinforcement of otherness, and how disability is arguably the clearest emblem of that alterity. In relation to research in medical science, disability plays a paradoxical role: on the one hand being central to justifying the projects of biomedicine and medical genetics, as it is used to garner public support for research by emphasizing the tragic nature of disability; and on the other being marginalized, as disabled people are discussed rather than listened to. Moreover, genetic research is closely linked to particular conceptions of normality: 'we can be certain', write Brock, Doerfler and Ulrich, 'that scientists are not currently searching for a "democracy gene" because the belief in democracy and capitalism are assumed by practitioners of modern science and are not considered a disease' (p. 149). There might be searches for the 'gay' gene or the 'God' gene, or for other forms of social deviance, but not for the scientific materialist gene or the secular pluralist gene (let alone, as one cartoonist has it, for the gene 'that makes you believe anything scientists tell you'). This tendency at least in some scientific quarters to search for an implicit genetic validation of certain conceptions of true humanity arguably extends to the concept of normality itself, in the problematic assumption that it is possible to distinguish genetic mutation from genetic variation, and in doing so already to render the former a 'biological defect' and therefore a suitable candidate for therapeutic intervention. This in turn readily melds with the supposition that sickness is a predicate not of particular individuals but of strands within the nebulous collectivity that is the human 'gene-pool'. Not only is the emphasis then placed on the search for genetic solutions (and its frequent corollary, the prenatal elimination of individuals that are bearers of biological mutations) rather than the alleviation of sickness through environmental means; it also suggests, through the metaphors it evokes, a matrix of thought within which certain human beings might be segregated into second-class membership of the human species by virtue of their genetic heritage.

When we look at the intellectual and cultural milieu, the background of attitudes and practices that lie behind prenatal testing, it becomes evident therefore that the issue of limitations on resources is bound up not only with discriminatory attitudes but also with a set of assumptions about the concept of normality that are intrinsic to the organizational categories of genetic science. Yet, when all of this is connected with parental desires for healthy (that is, non-disabled) children, this suggests an ironic inversion of the tension between the pressures towards expanding health-care and the limits on the financial and other resources to meet these. For while postnatally the needs of a child with disabilities and financial constraints may be in conflict, at the point of prenatal testing the parental desire for a healthy child and the limits of the health-care budget may operate to reinforce each other. Parental fears about bringing up a disabled child, discriminatory social and professional attitudes, the confines of governmental or health insurance funds, and perhaps even the basic metaphors of genetic science, may all conspire together against the chances of life for an unborn child that tests positive. Indeed, given the force of these other factors, it may not be unreasonable to suppose that ultimately what is at issue is not a question of resources at all, but a certain set of assumptions which serve to define in advance what priorities should be made in allocating those resources.

The question of eugenics

It is not an idle question, therefore, whether contemporary practices of prenatal diagnosis amount to a form of eugenics. According to the standard popular account, as Amy Laura Hall reminds us, eugenics belongs to another era. It depended, we are told, on the mistaken scientific belief that the selective breeding of superior individuals and the discouragement of inferiors from reproducing would result in the biological improvement of the human species, and on this basis sanctioned coercive measures in the service of obnoxious racist and elitist goals, epitomized in the Nazi programmes of sterilization and euthanasia of people with hereditary and other disabilities. Current practices, by contrast, are based on no such pseudo-science, and involve no coercion: talk of 'consumer eugenics', '*laissez-faire* eugenics' and the like in relation to antenatal testing is inflammatory language that fails to capture its decisive differences from historical eugenics.

Yet this is somewhat disingenuous, as several pieces in this volume show. Tom Shakespeare, for example, argues in the person of the 'opponent' in his imaginary dialogue that while there may be no formal instruments of coercion, pregnant women do not in practice have entirely unconstrained choice in relation to their decisions leading up to and resulting from prenatal tests. The inadequacy of the information they receive about the disability diagnosed, including the perspectives of people who have that disability, the failure to provide counselling on the psychological effects of abortion or even to say that abortion may be the only 'therapy' on offer, the bias shown by medical professionals, the expectations created by the provision of a test, and increasing social intolerance of disability when one could have chosen not to have one's child, all militate against women continuing with affected pregnancies.[2] For Shakespeare, the alternative is to ensure that women's choice is genuinely informed, and to reject any decisions for abortion that are eugenically motivated, that is, animated by the thought that disabled people do not deserve to live or are too expensive for society to care for.

Mary Mahowald similarly differentiates between reasons for abortion. She is happier on etymological grounds to embrace the language of eugenics, since it is reasonable for morally acceptable preventative measures – such as taking folic acid to avoid neural tube defects – to be termed eugenic, and so the issue is one of distinguishing good eugenics like this from bad eugenics. For her, an example of bad eugenics would be the choosing of abortion on the grounds that the child will be disabled, since this implies that a single undesired trait is being allowed to justify the destruction of the whole of which it is a part. If, however, the decisive reason for the abortion is that a woman would be unable to provide care for a disabled child, and no other source of help is available, her choice for abortion would not be bad eugenics. Mahowald makes clear, however, that even if this might apply in the case of an individual woman, such a rationale can never obtain for societies as a whole, which at least in the Western world do have access to the necessary resources and so have no justification for pursuing policies of prenatal testing with a view to the abortion of affected foetuses.

It is worth noting in passing that Mahowald's argument can be rolled back and made into a justification for providing adequate resources for all women who wish to continue with pregnancies, not merely those who receive a positive antenatal test result: if society as a whole is not justified in refusing resources for women who wish to continue with affected pregnancies, no more is it justified in refusing those resources for women who are unable to provide adequate care after birth in cases where disability is not at issue. Such a conclusion could be heartily endorsed by those who do not share her views on the morality of abortion as such, but who like her recognize that women facing crisis pregnancies might well feel rather differently about their situation if they had the wherewithal to cope.

More to the point, however, is the question of how easily her distinction can be made. While one can think of examples where a woman might be able to distinguish her attitudes towards a child with physical or mental impairment from her fears about not being able to cope, equally one can also envisage situations in which a woman's internalization of negative social constructions of disability would make it almost impossible in her own mind to disentangle her own attitudes towards disability from her worries about providing adequate care. And even if somehow such a distinction could always be made and could be made operative in some meaningful way – that is, if a pregnancy could morally be terminated not for reason of hostility towards disabled people, but because parenting a child with an impairment would present a major practical and financial burden – the implied message would still be the same: namely, that disabled people are an inconvenient nuisance and allowed into society only on sufferance. In a society where disabled people still feel tolerated rather actively welcomed and affirmed, abortion connected with disability will continue to be experienced as a sign of rejection.

Shakespeare questions this conclusion. He notes that the 'expressivist argument', namely that termination of pregnancy on grounds of disability is perceived by disabled people as a negative judgement on them, can take consequentialist and non-consequentialist forms.[3] In its consequentialist version, the argument claims that prenatal diagnosis followed by termination is likely to lead to negative consequences for disabled people, perhaps through contributing to more discriminatory policies or social practices. In its non-consequentialist version, it claims that termination because of disability is wrong because of the discriminatory attitudes it conveys. But, argues Shakespeare, the consequentialist form of the argument disregards the evidence that selective termination of pregnancy has happened at the same time as social provision for disabled people has improved: the evidence is lacking that prenatal testing leads to greater discrimination against them. And it is difficult to give any sense to the non-consequentialist version of the argument: existing disabled people cannot be harmed by termination since they are already alive; nor can foetuses before a particular stage of development – say 24 weeks – since they are only potential persons, and therefore have no interests which could be harmed; while foetuses after the stage at which they should be regarded as being persons with interests and rights should not be aborted for reasons of disability – or indeed for any other reason. So long as terminations of pregnancy are limited to the period before that stage of development is reached, the interests of no actual persons are harmed by them, whatever the motivation for them may be.

Even if one were to accept Shakespeare's view that the foetus is not a person before (say) 24 weeks, implicitly to equate the disability-related termination of a foetus at (say) 20 weeks with a sugar-lump inoculation for polio, on the grounds that both prevent disability but neither is equivalent to killing, ought to be at least a little unsettling: can we really be *that* sure that the fluctuating and medically contingent point of viability will take the moral weight of a purported change in status of the foetus from an 'it' to a 'he' or a 'she'? Whatever the case may be here, his argument tends towards the downplaying of the significance of the feelings of those disabled people who find selective termination of pregnancy threatening. This might be justified philosophically in a number of ways: for example that even though their feelings might be hurt, their interests have not been harmed; or that their feelings are being given moral weight, but just not as much weight as the feelings of those prospective parents who do not feel able to cope with the demands bringing up a disabled child may bring. Yet whichever route is taken here, the hurt is likely to remain: the continuity between the mindset behind antenatal diagnostic procedures and the prejudice experienced in everyday life is liable to remain too potent for the memories to be easily repressed. Prenatal testing stands as an ever-present reminder that those born with congenital disabilities take their place in society not by right, but by having the luck to have parents who chose to keep them.

Nor is this especially diminished by ensuring that consent is fully informed at the time of considering abortion after a positive test result. It is of unquestionable value that those considering their options in such a situation are given all the appropriate information – what it might be like to live with a child with the particular disability, what perspectives disabled people have, what the psychological consequences of abortion might be, and so on – in order that they can have genuinely informed choice.[4] But equally it would be misleadingly sanguine to imagine that prejudicial attitudes will be entirely eradicated by education. An individual's attitudes towards disability are forged at a much deeper level of buried anxieties about his or her own vulnerability and encroaching mortality, and while these fears remain, disabled people will continue to be the objects of other people's projected lack of self-acceptance. Even with properly informed choice-making, people will continue to abort, and those with mental or physical impairments will continue to feel disabled by the dominant attitudes of society: they will still feel the fragility of their social identity.

Gift and providence

This sense that disabled people may be allowed to scramble onto the back of the bus – or may not be – before it leaves, betrays a failure of hospitality by society at a profound level. And it is this lack of hospitality and of mutual recognition which the theological contributions address in different ways.

Bernd Wannenwetsch puts some searching questions to what he calls the 'inclusion paradigm', according to which the category of personhood is taken for granted and its bounds benevolently extended (or else not extended, sometimes also in the name of love or compassion!) to those whose personhood may be in question, such as embryos or foetuses in the early stages of development, adults in the late stages of Alzheimer's disease, patients in a persistent vegetative state, or those with certain severe mental or physical disabilities. Drawing on the German Roman Catholic moral philosopher Robert Spaemann, Wannenwetsch argues that the act of definition in the case of personhood is not a merely cognitive and distanced process of categorization, but is an intrinsically self-engaging phenomenon in which one discovers oneself as a person at the moment of recognizing the other as a person. There is a mutuality involved in the recognition of personhood which implies an inescapable ethical dimension, such that the primary question is not 'Are they human?' but 'What does their coming among us require of us?' The issue is not one of ascribing rights, as if we were in the quasi-divine position of deciding whether or not to attribute value to someone, but of acknowledging inherent claims on us, taking cognizance of a new fixed point of reference around which we are to organize ourselves. For this reason, people with severe disabilities, rather than being at the margins of the language-game of personhood, are at its centre, for they clarify that human dignity is fundamentally a matter of the humanity that is summoned forth in us as we recognize that we belong together and are called to be with each other.

There are clear parallels here with the lessons drawn by Jeffrey Bishop from work in the phenomenological tradition. While in Edmund Husserl and Martin Heidegger the direction of intentionality is from the self to other, this is reversed in Emmanuel Levinas, such that it is now I that am constituted by the other. Yet although Levinas's talk of the face of the other powerfully illuminates attitudes towards the unseen foetus who as yet has no face (and also shows how much more difficult it is to choose abortion once the face has glimmered into sight), Levinasian ethics still does not succeed in individuating the other, and in particular is silent before the face of the other who is unable to engage in discourse. In relation to persons, Bishop argues, what is needed is a sense of prior givenness, of the person as icon rather than idol. This distinction, taken from Jean-Luc Marion, refers on the one hand to the idol-creating constitution of the object by the subject, which absorbs the broken body into the categories of scientific medicine and subordinates it to the demands of economic performance. In the image of the idol is thereby constructed the category of 'disability', which unites all the disparate instances of impairment not as forms of biological malfunction but as different species of economic non-productivity. By contrast, the icon comes over time to constitute us as subjects, and calls us into being as persons out of our 'objectifying stupor', enabling us to recognize the truth of the person that is before us. When the other is seen as icon, the moral spotlight turns on the subject, and asks of us how we will nurture what is being asked of us.

The culture of choice is questioned at its roots in Hans Reinders' chapter, which advances the discussion through a critique of contemporary accounts of the basis of goodness. It is a frequent refrain in liberal thought that the good life, the life worth living, perhaps even the only life worth living, is one that we would choose. As Reinders notes, though, this places parents of disabled children who claim to have found their experience profoundly rewarding in a quandary: if faced with the question, 'Would you have this child again?', either they answer that they would not, which casts doubt on how enriching their experience of parenting really was, or they insist that they would, which renders their credibility suspect. Yet the problem lies in the assumption behind the question, that the goodness of a life is contingent on our being willing to affirm it by choosing it. If one rejects the implicit voluntarism in this conception of goodness, what remains is the idea that a life might be good *as it is*, independently of whether or not it might be chosen. Such an acceptance of the goodness of life as we find it might be unfortunate if this was interpreted in terms of an ontological identification of being with goodness, at least if this is understood teleologically to imply that human being possesses goodness to the extent that its natural capacities have been realized; for this might suggest that those unable to develop natural human capacities to any significant degree can never possess fully realized human goodness. Much better, argues Reinders, is to talk not of the inert 'givenness' of the good but rather of the good as a 'gift'. This points beyond itself to a giver, its goodness being dependent on the goodness of that giver. And this in turn relies on learning to trust in divine providence, that 'whatever it is that happens in our lives, comes from a loving God'. Without such trust, there is only outrageous fortune; with it, we can begin to be disciplined into living without resentment, receiving and giving love as companions together in God's time.

In different ways, each of these theological contributions are attempting to express the foundational solidity and non-negotiable presence of disabled people's lives, and therefore of all human lives. The location of any individual's value beyond our choosing or willing allows us confidently to greet each other: 'it is *good* that you are here'. Good, not because you might turn out to be a rich source of blessing to me – for, equally, you might not. Good, not because your presence may enable me to work out some of my existential anxieties – for that also makes your existence instrumental to my needs. Good, not because your different giftings may make me think of you as especially lovable or freed from vanity or selfishness – for I may also find that your differences make you aggressive or unreceptive to my love. But good, simply because you are here, the gift of a loving God who welcomes us both.

Whether such unconditional delight in another is finally possible without an implicit trust in providence is an open question. After all, might it not be that without some such tacit sense of the individuated gratuity both of our own existence and that of others, we are all afloat on the sea of being measured by our capacity to fulfil certain potentialities? If I do not find you attractive or productive or worthwhile, and indeed may see you as the opposite, as a drain on my limited resources, what reason do I have for valuing you? But by the same token, is it not true of all of us that shorn of grace we only exist because we are treated as valuable by others, and therefore that our lives are contingent on their valuing us? Whatever may be the case here, we should not be surprised that without a fundamental affirmation of the non-contingency of their presence in society, that they are not finally to be measured against the stern rule of performance, disabled people will continue to feel profoundly vulnerable to suggestions that their lives are not worth living. And what is true of disabled people is finally true of all of us, disabled and non-disabled alike.

Brent Waters pushes one step further the idea that all human beings, disabled and non-disabled alike, share a common destiny. If we find in Jesus Christ the human image of the divine perfection to which we are called, disability cannot be properly understood as physical or mental impairment, but as anything which prevents us from being made perfect in Christ. Our sinfulness is therefore our true disability: all human beings are disabled, for all have sinned. In saying this, Waters is only too alert to the dangers that such expansive and inclusive redefinitions can render physical and mental impairment socially invisible, with the political consequences of yet again marginalizing the problems faced by disabled people. His concern, however, in recasting the issue in this way is not to deny the reality of impairment, but to underline that, before God, the common calling in Christ which disabled and non-disabled share with each other categorically transcends whatever differences they may have from each other. It also allows him to draw parallels between the temptation to escape from sin through Pelagian efforts at self-improvement, and the temptation to evade finitude and bodily frailty through equally Pelagian attempts at eugenic enhancement of the human race. The Early Church's condemnation of this ancient heresy was not about an abstract issue concerning the means of salvation, but described an all-too-concrete set of motivations which even in the twenty-first century are being expressed through particular cultural and medical practices with profoundly destructive consequences.

Embodiment in practice

Without embodied practice, fine words about disability and the lives of disabled people remain just so many words. The test of theological claims is not just a matter of reflective assessment of their intellectual consistency: if theology is an essentially self-engaging discipline, it must be intrinsically and not merely contingently related to Christian practice. In a world in which we are taught that we may see embodied expressions of resurrection life, we should expect to see such claims exemplified.

This explains the significance of stories such as that recounted in Reinders's chapter about Oliver de Vinck, whose profound disability drew so much out of his family. In reading this account it is important to see it not just as an anecdote, as if it were in a second-class category of social scientific data, nor as an incidental story designed to illustrate a deeper and more 'real' intellectual point, but rather as itself an iconic illumination, in which the presence of the resurrected life comes to visibility. Here we find a brother who in recognizing his brother not as a 'vegetable' but simply as Oliver, found his own deeper humanity. Here we find a mother who by not clutching onto her plans for her life, and by learning not to resent her circumstances, eventually discovered in her son's life a blessing that would renew her, making her 'more consciously, more powerfully alive'. Here we find a family who refused to think of a severely disabled boy in terms of a choice, and discovered in him a gift. Here perhaps we even find some incipient response to the economics of scarcity, in which the question was not the provision of food for the next 32 years, but sufficient provisions for today, in which resources of time, attention and love were drawn out of Oliver's family which they might never have had access to otherwise.

Of course there are dangers in elevating Oliver's family to stained-glass sanctity, as they would know only too well. Nor should we forget that books in this genre of living with disability are likely to be written by those who have emerged triumphant, not by those whose mental health has been severely compromised on the way or whose relationships have broken up or careers been destroyed as a result of the experience of disability. Nor can the risk be entirely evaded of working out one's theology in theory and then claiming to find it embodied here in life. Yet for all this, maybe there is still a chance that stories such as that of the de Vincks may be able to speak as a living witness to a different story about disability than that given in the dominant cultural narratives – that is of disability as tragedy, indeed as a tragedy which medical science may help vigilant parents avoid.

For Christians this different story is inseparable from the one that has been heard and is told by the Church through its liturgies of word and sacrament. But if the thrust of the theological contributions to this book is to be believed, and an understanding of disability is to be placed at the centre of our understanding of humanity rather than at its margins, then it follows, as Martina Holder-Franz points out, that disabled people should be at the centre of our liturgies, not as an ideal but as an everyday reality. While no doubt a further intellectual agenda arises out of this volume, theology's primary witness must be that humanity will never find salvation apart from coming to participate in Christ's own broken body.

Notes

1. At least in the UK: for an analysis of a range of social-contextual models of disability, and a critique of the British strong social model, see Tom Shakespeare, *Disability Rights and Wrongs* (London: Routledge, 2006), pp. 9–82.
2. For a collection of stories told by women who have defied these medical and social pressures, see Melinda Tankard Reist, *Defiant Birth: Women Who Resist Medical Eugenics* (North Melbourne, Victoria: Spinifex Press, 2006).
3. Here I also refer to Shakespeare's more extensive account of his position in *Disability Rights and Wrongs*, pp. 85–102.
4. See further the ANSWER website at www.antenataltesting.info.

List of Contributors

Jeffrey Bishop is a doctor currently working as a senior lecturer in medical ethics and law at Peninsula College of Medicine and Dentistry, University of Plymouth, Devon.

Brian Brock is lecturer in moral and practical theology in the School of Divinity, History and Philosophy at Aberdeen University. His particular interests are the ethics of new technology, the use of the Bible in Christian ethics, disability and political theology, and medical ethics.

Stephanie Brock is a former nurse and fieldwork researcher in the field of disability. She is currently working in an advisory capacity for the Centre for Spirituality, Health and Disability at Aberdeen University (www.abdn.ac.uk/cshad). Stephanie is married to Brian Brock and is the mother of Adam.

Professor Walter Doerfler is an internationally renowned virologist and specialist on adenovirus. He currently heads the department of Medical Genetics and Virology at the Institute for Genetics in the University of Cologne.

Amy Laura Hall is Associate Professor of Christian Ethics and director of the Doctor of Theology program at Duke University Divinity School, North Carolina. She has written two books and many articles, and received grants from the Henry Luce Foundation and the Lilly Foundation.

Revd Martina Holder-Franz was born and raised in the former GDR She worked for four years with children and adults with mental and physical disabilities and has a diploma in social education. After theological studies in Basel, Berne-and Marburg she is now a pastor in the Swiss Reformed Church. She has special training in systemic counselling and is at present working on her dissertation, an interdisciplinary study in theological ethics and practical theology.

Mary Mahowald is Professor Emerita in the Department of Obstetrics and Gynaecology, Committee on Genetics, and MacLean Center for Biomedical Ethics at the University of Chicago.

Christopher Newell is Associate Professor of Medical Ethics in the School of Medicine at the University of Tasmania, Australia. He is associate priest at St David's Anglican Cathedral, Hobart, Tasmania. Dr Newell is a person with impairments who has long drawn on-his experience of disability and mortality to explore the ethical and spiritual dimensions of life. His awards include being made a member of the Order of Australia (AM) in 2001.

Hans Reinders is professor of ethics in the Faculty of Theology at the Vrije University in Amsterdam, where he also holds the Bernard Lievegoed Chair for

Ethics and Disability. Professor Reinders has published on a variety of topics related to intellectual disability, including the effects of the expanding practices of genetic screening and testing for people with disabilities.

Tom Shakespeare is the director of outreach for the Policy, Ethics and Life Sciences Research Institute, a Newcastle-based project developing research and debate on the social and ethical implications of the new genetics (http://www.ncl.ac.uk/peals/).

Blair Smith is professor of Primary Care Medicine at Aberdeen University. He has published widely in the fields of genetic epidemiology and chronic pain, and he is national coordinator of Generation Scotland: the Scottish Family Health Study. A member of the scientific management group of Generation Scotland, a multidisciplinary research organization, he works to further its aim to create an ethically sound, family and population-based infrastructure to identify the genetic basis of common complex diseases.

Robert Song is senior lecturer in Christian Ethics at Durham University. He is also a member of the Scottish Regional Collaborating Centre for UK Biobank. He is author of *Christianity and Liberal Society* (Oxford: The Clarendon Press, 1997) and *Human Genetics: Fabricating the Future* (London: Darton, Longman & Todd, 2002), and many articles on ethics. His current research is on a theological narration of the development of philosophical bioethics.

John Swinton is professor in Practical Theology and Pastoral Care in the School of Divinity, Religious Studies and Philosophy at Aberdeen University. He has a background in nursing and mental health chaplaincy and has researched and published extensively within the area of mental health, spirituality and human well-being and the theology of disability. He is the director of the university's Centre for Spirituality, Health and Disability. (www.abdn.ac.uk/CSHAD).

Hans Ulrich is professor of theology at Erlangen-Nuremberg University.

Bernd Wannenwetsch lectures in moral and systematic theology at the University of Oxford. He has written widely in the fields of ethics, political theology and the theologies of Luther and Dietrich Bonhoeffer.

Brent Waters is director of the L. Jerre and Mary Joy Stead Center for Ethics and Values, and Associate Professor of Christian Social Ethics at Garrett Evangelical Theological Seminary, Evanston, Illinois. He has served previously as the director of the Center for Business, Religion and Public Life, Pittsburgh Theological Seminary, and has written widely on ethics and medicine.

Index

"ableism" 109
abortion 2–7, 20, 31–2, 35, 40, 50, 59–60, 67–70, 78, 92, 97–100, 137, 163, 207, 219, 229, 236–41
    disability rights critique of 102–3
    after diagnosis of Down's syndrome 104–6
agape 46
ageing 206
agency 48
alcoholism 48
Alzheimer's disease 10, 97, 189, 201, 241
American Breeders Association 91
American Eugenics Society 76–80, 84–5, 90–1
amniocentesis 4–7, 104
angels and the angelic mission of the disbled 192–7
angina 135
Aquinas, Thomas 171–3
Arendt, Hannah 190
Aristotle (and Aristotelianism) 100, 203, 216
Asch, Adrienne 102, 106
association studies 139
Augustine, St (and Augustinian tradition) 40–1, 171–2, 203–4, 229
autonomy 52–3, 92, 97, 100, 105, 207–8

'Baconian project' 170
Barth, Karl 40
being, goodness as 171–4
Berube, Michael 38
bioethics 45–7, 163–70, 220–1
biomedicine 119–25, 219–21, 236–7
    disability rights critique of 214
    implications of molecular genetics for 126–7
biopolitics 218–19

biotechnology 53–4, 205
    see also genetic technology
Bishop, Edwin 84–5
Bishop, Jeffrey P. 23, 241, 245
    author of Chapter 13
body, the: definition of 47
    objectification of 214
Boethius 187
Bonhoeffer, Dietrich 40–1, 189
bowel cancer 134
brain function 117–19, 130
breast cancer 97, 134–8
Brent Waters *author of Chapter 12*
Brock, Brian 12, 17–23, 236–7, 245
    co-author of Chapter 1, co-author of Chapter 9 and co-editor
Brock, Stephanie 20–1, 236
    co author of Chapter 1
Brueggemann, Walter 1, 15, 23
Buber, Martin 62
Buck v *Bell* case 76–7, 96–105
Burbank, Luther 86–7

Calvin, John (and Calvinism) 204
Canguilhem, G. 218
care: of infants 103, 239
    of those with Down's syndrome 104–5
    see also health care
Carruth, William Herbert 86
categorical imperative 223
China 70
choice as the necessary condition for life's goodness 168–71, 174–7, 241–2
cholesterol 135
Christian Century 80
chromosomal abnormalities 123
civic duty 81, 90–1
cleft lip and palate 97
cloning, *reproductive* and *therapeutic* 121,

Index

125, 128, 158
cochlear implants 210
Cold Spring Harbor Laboratory 76–7
commodification of children 9–10
conditio humana 117
contemplative life 11, 203
contraception 99–102
Cook, Michael 52
cost-benefit calculations 70
counselling 97
   see also genetic counselling
Crick, Francis 133
cystic fibrosis 6, 13, 48, 50, 97, 134, 136, 201, 206

Darwin, Charles 60, 78, 83, 153, 155
Dasein 222
Davenport, Charles 76–80, 90
Davis, Lennard J. 45, 49
de Vinck, Christopher 174–8, 243–4
deaf culture 49
Dealey, James 89–90
dementia 10
   see also Alzheimer's disease
diabetes 134
dignity, human 189, 241
Diprose, Rosalyn 48
disability: concept of 202
   and connotation of inferiority 205
   disparate features encompassed by term 219
   elimination of 165, 235
   moral encoding of 44–5
   public attitudes to 68, 70, 207, 238, 240
   as sinfulness 210, 243
disability rights movement 45, 221, 236
   critique of medical practice by 98, 102–6, 214
discrimination 49, 68, 70, 103, 166, 202, 237–9
distributive justice 92
divine liturgy 229–30, 244
divine providence, doctrine of 13, 177–8, 242
DNA (deoxyribonucleic acid) 48–9, 118–20, 124–5, 130, 133–5, 151, 234
Doerfler, Walter 17–18, 22, 153, 234–7, 245
   author of Chapter 7 and co-author of Chapter 9
Dolly the sheep 127–8
Donaldson, George Huntingdon 84–8

Dorris, C.L. 89–91
Dostoevsky, Fyodor 222
Down's syndrome 4–7, 13, 21, 29, 33–8, 50, 59, 97–8, 102–3, 136, 155–6, 166–7, 201, 206, 211, 236
   prenatal testing and selective termination for 104–6
drug-dependency 48
Durst, Dennis 75
Dworkin, Ronald 69

Easton, David 30
economics 218
Edwards, Jonathan 85
Eliot, Charles W. 89
Ellis, Henry Havelock 89–90
Ellwood, Charles A. 89, 91
embarrassment about disabled people 192
embryonic stem-cell research 53–5, 121, 127–8
end-of-life issues 169
Engelhardt, H. Tristam 170
Enlightenment ideology 53
enucleation 29
epigenetics 120–1, 125, 128, 151
'epiphany of the other person' (Levinas) 62–3
ethics 133, 146–56, 214–15, 225
   genetic 149–51
   see also bioethics
ethics committees 52
Eucharist 211
eugenics 3–6, 23, 50–1, 60–1, 67, 70–1, 75–92, 96, 129, 163, 165, 205–10, 214, 220, 238, 243
   coercive and *voluntary* 91
   definitions of 98–9
   good and *bad* types of 3–4, 51, 97–105, 206–9, 238
   liberal 206–9
   as a whole spectrum of concepts 98–101
Eugenics: A Journal of Race Betterment 81–2
euthanasia 8, 206, 229, 238
evolution, theory of 152–4, 209
expressivism 7
'expressivist argument', the 102, 106, 239

family histories of genetic conditions 70
Fletcher, Joseph 45–7
flourishing, human 209
'foetal interests' 69
foetal monitoring 201

248

Index

foetuses, legal status of 105
folic acid 238
foot-binding 82
Foucault, Michel 47, 216–20, 226
French Revolution 217–18

Galton, Francis 87, 99
gene discovery studies 139
gene-environment interactions 140
gene expression 151, 155
gene therapy 121, 126–7
genetic counselling 102
genetic disorders 122–5, 154–5, 163–5, 234
genetic dispositions, information on 166–8
genetic engineering 76, 205
genetic enhancement 206–7
genetic epidemiology 134–6
　listing of studies 141–3
　study types 139–40
genetic information 48, 236
genetic risk 138
genetic science 17–18
　aims of 132, 235
　and clinical practice 134–5
　limitations of 150–1
　and need for the Church 14–15, 22
genetic studies, population-based 136
genetic technology 2–8, 12–13, 17–22, 30, 120
　see also biotechnology
Geneva Convention 190
genocide 99, 101
Germany 128–9
Gilbert, Walter 1, 3
Gilson, Etienne 171, 173
God: authority of 16, 20, 22
　faith in 64
　see also divine providence
God's love 11–12, 18–19, 39
Goggin, G. 53
'good life' concept 163–71, 177–8, 242
goodness as being 171–4
grace 210–11, 242
Grant, George 29–30

Habermas, Jürgen 207–9
haemophilia 201
Haldane, J.S. 80–1
Hall, Amy Laura 3–4, 20, 23, 238, 245
　author of Chapter 5
happiness 9, 117
Harris, J. 220

Hauerwas, H. 21, 30, 40
'having a life' as distinct from 'being alive' 170
health, definition of 171
health care 132, 235–7
　primary 138–9
health-illness spectrum 132
'healthy unwell' 49
Hegel, G.W.F. 83
Heidegger, Martin 219–22, 225–7, 241
heredity 86
heritability 137, 140
The Hitch-Hiker's Guide to the Galaxy 52
HIV 39
Hockenberry, John 54
Holder-Franz, Martina 6, 22–3, 244–5;
　author of Chapter 3
Holmes, Oliver Wendell 76–7, 96–101, 103, 106
Holocaust, the 3
Holtzman, N.A. 49
Homiletic Review 76, 87, 89
Hubbard, R. 48–9
Human Genome Project 2–3, 50, 133
human nature 117
humanness 46, 55, 132
Huntington's chorea 13, 97, 136–7, 152
Husserl, Edmund 222, 225–7, 241

imprinting defects 123–4
Incarnation, the 203, 210
'inclusion' 182–3, 241
intellectual disability 5, 164–5, 170, 178
in vitro fertilization (IVF) 201, 207
ischemic heart disease (IHD) 135–6
Ivory soap 91–2

Jesus Christ 63–4, 86–8, 203, 209–12
Johnson, Mary 50

Kantianism 220, 223, 228–9
Kingsley, Charles 80
kinship with the disabled 189–92
Krakauer, Eric 214, 217–18, 227

Ladies' Home Journal 91
language, use of 146–7, 151–7
Leon, Sharon 75
Levinas, Emmanuel 62–3, 215, 221–7, 241
liberalism, political 20, 92
Lindemann Nelson, Hilde 47–8
linkage studies of gene loci 139
Lombardo, Paul 96, 98, 101

Index

love 63–4
  see also God's love
Luther, Martin 39–41
Lynn, Richard 51–2

McCulloch, Oscar 81
Machiavelli, Niccolò 47
Macintyre, Alasdair 53, 168
McKenny, Gerald P. 170
Mahowald, Mary 4, 23, 238–9, 245
  author of Chapter 6
Marion, Jean-Luc 215, 224–9, 241
marriage 89–90
Marteau, T. 49
Mather, Cotton 204
Mendel, Gregor 132, 204, 234
mental disability 62, 92
mental retardation 104–6, 164
messenger RNA 119
metaphysics 216
Methodist Quarterly Review 84, 87–91
Mettner, Dieter 60–1
Micklos, David 76–7
Milbank, John 224
mine-clearance 69
mitochondrial DNA 124
Mitwelt concept 182, 185–6, 195
molecular genetics and molecular medicine 119–22, 126–7, 234
monogenic diseases 123
Morrow, Derek 229
Mote, Edward 86
motor neuron disease 13
multifactorial diseases 123
mutation 124, 149–56, 237
myocardial infarction 135

Nazism 3–4, 67, 76–7, 96, 99, 101, 105, 238
Newell, Christopher 4, 22–3, 53, 230, 236, 245
  author of Chapter 2
normality, concept of 60–1, 202, 237
nucleotide sequences 121

obesity 209–10
O'Donovan, Oliver 153
Oë, Kenzaburo 194–5
original sin, doctrine of 203
Osborn, Henry Fairfield 87
Osgood, Phillips Endecott 76, 84–8
osteoporosis 134

other-regardingness 168
ovarian cancer 134

Parens, Erik 102
parent-child relationship 207
Parsons, Philip Archibald 89
Passmore, John 202–5, 209–10
Paul, St 13, 15, 211
Paul, Diane 98–9
Pelagius and the Pelagian heresy 203–4, 209–10, 243
perfection, quest for 60, 102, 165, 202–5, 209–10
persistent vegetative state 241
personalized medicine 136
personhood 8–12, 46, 48, 61, 182–96, 208–9, 241
pharmacogenetics 137
phenomenology 222–9
phenylketonuria 122, 134
  see also Tay-Sachs disease
Pieper, Josef 19, 21
Plato 203
'playing God' 165
Plessner, Helmuth 182–6, 193–5
Polanyi, Michael 190
'pool' of all possible knowledge 118, 130
poor, the, mission of 188
Popenoe, Paul 76, 78
populism 90
preference utilitarianism 8
pregnancy, termination of *see* abortion
pre-implantation genetic diagnosis (PGD) 201–2, 206–7, 210
prenatal testing 2, 4, 7, 30, 32, 20, 58, 60–1, 67–8, 92, 97–102, 163, 166, 236–40
  disability rights critique of 102–4
  for Down's syndrome 104–6
'presence of peace' 174–7
privacy 166, 168
'privatization of the good' (Macintyre) 168
'problem' of disability 1
'procreative liberty' 206–7
progressivism 83
propaganda 90
prosthetic limbs 201–2, 210
protein structure 119
Protestantism 78–81, 87, 90–2, 204
Psalm 22 12–13
Psalm 127 39
Psalm 139 12

psychological aspects of medicine 118
'quality control' 9, 206–7
quality of life 61–3, 167, 205
  see also 'good life' concept

racism 78, 88, 238
Ramsay, Paul 46–7
Reeve, Christopher 52
Reiland, Karl 80–3
Reinders, Hans S. 10–11, 20–3, 92, 194, 241–3, 245
  author of Chapter 9
Religion in Life (journal) 88
resource-oriented view of life 62
resurrection of the body 212
ribosomes 119
risk assessment 128
road safety 69
Roman Catholic Church 80
Rosen, Christine 75, 80–5
Rowe, Gilbert T. 88–9
Ryan, John A. 82–4

Scotland 37
self-regardingness 168
sermon contests 84–5
Shakespeare, Tom 4, 23, 238–40, 246
  author of Chapter 4
sickle cell diseases 97, 134
Singer, Peter 8–10, 183, 189
Smith, Blair 17, 22, 234–5, 246
  author of Chapter 8
smoking 135
social attitudes 106
social construction 102, 202
social Darwinism 48, 60
social engineering 68
social policy 52
social welfare 100
socialization 208
Song, Robert 23
  author of Conclusion 234, 246
sonograms 31, 221
Sonoma State Home 76
soul, human 29
Southard, Helen 76
Spaemann, Robert 182–8, 241
speciesism 189
Spencer, Herbert 78

spina bifida 97, 219
stem-cell research *see* embryonic stem-cell research
sterilization 77, 83–5, 88–91, 96–101, 238
Stinkes, Ursula 60
Strong, Josiah 89
'survival of the fittest' 90
Swinton, John 246
  author of Introduction and co-editor

Tay-Sachs disease 97;
  see also phenylketonuria
Taylor Sumner, Walter 80
Tennant, F.R. 204
testing, ideology of 35–40
  see also prenatal testing
Thatcherism 53
Thomism 171
trinucleotide repeats 124

UK Biobank 136
Ulrich, Hans 17–18, 236–7, 246
  co-author of Chapter 9
United States 78–9, 210
  Supreme Court 76, 96, 101
utilitarianism 8–9, 46–7, 50, 218

vaccination 69
Vanier, Jean 63
Verhey, Allen 2–3, 10
Vietnam War 30

Waddington, C.H. 120
Wald, E. 48–9
Wannenwetsch, Bernd 10–11, 23, 241, 246
  author of Chapter 11
Ward, Harry F. 82–3
Ward, Linda 2
Waters, Brent 9, 243, 246
  author of Chapter 12
Watson, James 76, 133
Wells, David 53
Winship, A.E. 85
'worried well' 138
worship 13, 211

Xenophon 85